Utopia Rising

Matthew L Sexton

ISBN: 1-4635-0884-0
ISBN-13: 9781463508845

Dedication

For those who value humanity's progress

and dream of our ultimate potential.

Table of Contents

Special thanks to CreateSpace for their services, to my aunt, Norma Nelson, for her invaluable assistance, and to my parents for their support.

Introduction

This book is meant to be a general outline of my social philosophy. Though social philosophy can be said to encompass all political, economic, ethical, and judicial topics, I have attempted to limit the scope of this work to that of purely societal matters. However, to accurately define them, you must touch on these other aspects to some degree since they are interrelated. Society interacts significantly with politics in order to function. Outlining an entire political system in detail is beyond this current objective, although I do hope to approach these other facets of civilization in due time and with due consideration. Ethics and moralities will be purposefully avoided herein except in vaguely defining their sphere, their proper place in society. I have devoted a large portion of the material to my interpretation of the government's correct functions since all practical concerns not governmental in nature must ultimately fall into society's lap.

I have attempted to write this in such a way as to begin with the broad aspects and then to gradually move toward the points of greater precision. To generalize this flow would be like first validating your assertion that murder should be a crime before attempting to substantiate a particular view on whether or not it should be considered a capital crime. We tend to vary on the latter much more than the former. Hopefully the same holds true of my social philosophy. It is not terribly complex, and could possibly even be better defined as a point of view rather than a philosophy. Regardless, this simple outlook on society's progression has powerfully affected my views on political, social, and judicial matters. I continually find myself asking questions such as: Where will this eventually lead? What will be of greatest benefit to distantly future generations? What are our ultimate objectives?

What is a philosophy? In searching various dictionaries, we find that a philosophy is generally defined as a set of guiding principles and the study of how to develop them. In doing this, philosophies often explore topics such as the meaning of existence, the exact definition of reality, the nature of consciousness, and so on. Even at a glance one can see that these are very deep waters. One definition, from *The New Oxford English Dictionary*, reads: "A theory or attitude that acts as a guiding principle for behavior." This definition taken in a social context describes my focus for the work; this is what you're in for if you read on.

Since the manner in which we define words often changes over time and reflects certain aspects of society's views, it is interesting to note that among these various sources, two (*Merriam-Webster* and *Encarta*) imply that philosophy is calmness, rationality, and judgment. Would this mean that people now see philosophies worthy of consideration as generally drawing civilized conclusions? Can conclusions be substantiated without careful deliberation and scrutiny free from emotional biases? Social and political philosophies take significantly particular views on violence. You could say that their views regarding violence do more to define them than several of their more apparent aspects. What we have generally come to accept regarding the nature of violence in society suggests much about where we are going. It is the contrast of our views on civility. This violence/civility gauge of popular opinion speaks volumes about humanity and its potentialities. What your philosophy entails greatly defines your character. To clearly define yourself, you must define your views. They form your point of reference. Your actions are the other significant point in the equation.

Humanity's Fate

Where are we going? Is the world progressively improving or worsening? Everyone can agree that times are changing, as they always have, but are they changing for the better? Ask yourself this question. What would you answer? How would your acquaintances respond? How many, when asked where the world is going, would say, "To hell in a handbasket"? Is mediocrity our fate? Would any of them reply that the human world is destined to become a paradise?

This question is far from new. To quote Voltaire, "Mankind has always maintained that the good old times were much better than the present day." What was so wonderful about the past, and even the ancient past, that causes people to latch on to it so? Some say we think of the past as being better than the present because, in it, we were all younger. This can certainly make it seem as though the past will always be better than the future. What about the times, though? Look beyond your situation and reality. What about society, the environment we create collectively? Were the times always better further and further back in history? Should we seek to turn back the hands of time? If so, how far should we run back the clock? If society is progressively worsening, then the best of times must have been the dawn of man. Why should we covet the lifestyles of our primitive ancestors? Once we achieve this animalistic state, should we forbid discovery in the interest of keeping our lives simple? Should progress and invention be made into crimes against this creed of stagnation? There exists the assertion that people are at their best when times are at their worst, more caring for each other and so forth. Even taking this observation into account, is it a good enough reason to purposefully create difficult times just so that we can help each other get through them? Can you imagine a better world?

What defines the successful society? The most important factors I see are civility, safety, prosperity, knowledge, health, technology, independence, comfort, efficiency in dealing with societal and political issues, and the preoccupations of the collective will. Many of these aspects are directly related. In fact, you could combine them all as different forms of safety. These hallmarks should be our focal points. They are what must be improved if we are to better society.

- Civility is the desire for peace and, in my view, ultimately establishes unity. One way to measure it is by the lack of crime in society. Those who have learned to value its gains will naturally prefer an environment of cooperation. The desire for civility is influenced by the efficacy of the other factors, an appreciation of what it can facilitate, and/or a clear picture of what its opposite state can bring.

- Safety means security from all harm, both natural and man-made.

- Prosperity means strength of economy (again you could say safety from poverty).

- Knowledge is how deep you can dig for information in your environment. Imagine a library of ancient times compared to a modern library, or a class in any subject you enjoy three hundred years ago compared to now.

The books get more numerous and voluminous with time. The storehouse of knowledge requires addition after addition.

- Health is enabled by technology, knowledge, and the preoccupations of the collective will. In our world, what once would surely have developed into a cata-strophic health crisis is now prevented through easily achievable means.

- Technology determines our survivability and capabil-ity. If a natural disaster were imminent, how could we effectively face it without knowledge and technology?

- Independence, this is a large and widely debatable subject. How much freedom should we have? To what degree does our government need to threaten and set boundaries in order to achieve the best possible society and environment? Each and every law on the books is a trade-off between freedom on the one hand and the perceived benefits to the general well-being of society and the establishment of safety and justice on the other. Regardless of where you think these lines should be drawn, certainly we must have some degree of self-rule, some degree of independence, to function as a successful society. What kind of society only functions well when operated as a penal colony? Our capacity for self-rule has greatly increased from ancient times. Governmental structures are much more intricate and developed. Whatever degree of in-dependence you see as being ideal, the fact remains that there is a specific measure that will bring about the best results. I think this ideal amount (whatever it may be) will prove, in time, to be ideal for the entire world and not only for individual nations. It is our job to find where these optimal lines are to be drawn. The extremes of worldwide enslavement or pure anarchy would be far from ideal.

- Overall comfort is not only an indicator of a successful society, but is also facilitated by it. Comfort is bred by the advancement of the other factors in question. It is only possible in the right kind of environment. Economic prosperity alone is not enough to facilitate the comfort required for people to produce or improve to the best of their ability. The other hallmarks of a successful society are needed if people are to lead fulfilling lives. The average person's potentiality for achieving comfort speaks volumes about our progress.

- For a society to function well, a proven methodology for dealing with problems must be developed. If a question arises and no one knows what action to take or how to approach the situation, the society would not be said to be functioning very well. This aspect can be simplified as "the efficiency of the government" and "the efficiency of society." It is important for us to make the distinction between the two entities that rely on each other to create civilization. Into which sphere an issue belongs is one of our most problematic preoccupations. One of Utopia's most indicative aspects is that there is nearly always a fast, efficient, and effective response to whatever difficulties may arise. Only the completely unforeseeable problems are faced with a hesitation over how best to proceed once a proven methodology has been developed. In short, effectiveness of protocol exemplifies preparedness.

- Finally, the preoccupations of the collective will, what are the aims of the majority? What are society's goals and chief concerns? What big aims exist, aims no individual can achieve, but that the majority can accomplish? What aesthetic societal values will bring us closer to an environment that accurately displays our progress and potential? For any of these aspects to improve, there must be some degree of organization, drive, intelligent leadership, and cooperation. Unity can determine whether the goal is met or lost, whether a nation rises or falls. It can make or break the cause at hand.

Considering a time line of notable inventions, we can easily see that technology is obviously on the rise, but what about the other factors we could say define our species and its progress? What about peace, unity, prosperity, general living conditions, crime, health, and so forth? Obviously these factors take large steps forward and backward over time. The progress of peace looks grim during a world war. Prosperity looks grim during a depression. Crime looks as if it's on the rise during a riot. The trees look huge when you are standing a foot away from them, but what about the forest, the big picture? What can we do to facilitate the needed changes? Do you see an obvious problem with an obvious solution? What needs to be done? What stands in the way? We must remove the barriers to our progress thoughtfully, theorize and consider all points of view, put our ideas to test, and keep a sharp eye on our steps to spot our errors as quickly as possible. We must make great attempts to see our world refined. We must set goals and keep them clearly in mind.

The main assertion I wish to make is that, in my view, when looking at history from ancient times to the present, there can be seen an obvious trend in the direction of progress regarding each of these aspects of our world. We are becoming more civilized on the whole. We are becoming more developed, more prosperous, efficient, and healthy. We are gradually mastering our environments rather than being defenselessly subjected to them. The evidence for this observation is clear and abundant. Its origins are worthy of study at the biological, as well as the social level. Nurturing this instinct for betterment should be the most important of all human endeavors and our chief concern in life. The conclusion I draw from this continual progression is that it will, in time, lead to an ideal society relative to our own and that it will then continue to refine and improve our lives as it always has. With each step taken, we should ask ourselves if we are paving the way for the future we seek or if we are slowing its progress. If we have no clear picture of where we want to go, how can we get there effectively? The recourse is to wander about hoping to get lucky and hoping to retain progress without effort. Thus far, numerous improvements have been made to our quality of life without notice and many without conscious attention to this ultimate goal. Each invention, discovery, and popularly accepted societal improvement has nudged us along.

The second assertion is that this coming Utopia is inevitable. Humanity's extinction is possible, but its continual degeneration is not. If humanity progressively eroded, it would not be humanity as we know it. It would be something else entirely. In time, regardless of how long we allow a particular ideology or system to harm progress, we will eventually seek betterment and build on past achievements. Whatever else humanity's distant future may be, short of a catastrophe or massive setback, it will continue to improve and become more potentially fulfilling. This drive for betterment will continue so long as life continues. The desire for improvement is an intrinsic part of our nature. It is linked to our instinct for survival and is undoubtedly the greatest of all life's gifts. To ignore this drive for betterment is to ignore our potential, a potential that can theoretically never be completely fulfilled since there will always be room for improvement.

Anyone can argue against this idea that the future looks bright by pointing to any of the recent incidents (there are always plenty to choose from) that would seem to indicate the contrary. It is easy to be swayed by the tragic disappointments bombarding us on a daily basis. There are hundreds of things we could be doing much better. Many on this planet of ours are still set on killing each other while children are dying needlessly. This situation is unconscionable considering the progress made thus far. Still present are the same old indicators that existed in ancient times of mankind being barbaric and savage.

The good news is that these glaring indicators are fading. Not fast enough, but they are fading. Our world's society today is less tolerant of uncivilized behavior than ever before. Consider that the barbaric gladiator contests of old would never be allowed in today's world, or that executions have, over time, gone from being as inhumane as possible to as humane as possible. A genocidal dictator like Genghis Khan could not commit such atrocities in the modern world and expect no response from the international community. Atrocities have been forced into secrecy by a higher degree of global civility. This indicates that we are adjusting to our newfound technological capabilities with a certain degree of maturity. Once considered routine and common throughout every culture, murderous tyrants hiding behind the protection of their governments are fading fast. It would appear that their days are numbered. They have no place in the future. Uncivilized methods are becoming more and more useless in the face of human achievement.

Those who are most eager to jump to violence, the thugs, bullies, extremists, racists, tyrants, and warmongers of the world, all see the scientists, doctors, and the greatest minds of civilization with the entirely wrong attitude. They see a kind of intimidation and bitter jealousy, which, if examined closely, they would find is actually self-hatred. They see mankind's progress as a threat to their own obviously corrupt goals. This jealousy is over the fact that others are able to excel without using threats, prejudices, or violence, which is all these sad individuals know how to use to get what they want and to justify their various maleficent positions. They lack the creativity to do anything except destroy and steal, but they cannot destroy the masses' instinct for betterment, and as others have noted, time and again, they cannot steal the thought processes of great minds for themselves. Though they may kill more of the civilized than the civilized kill of them, two things stand out as a striking contrast: One is that the numbers of the civilized continues to rise while the numbers of barbaric brutes continues to fall. The other point is that they are inherently doomed to failure since, even if they were to conquer civilization, they would then destroy each other.

What are the ultimate objectives? This question is valid. The emotion we should all feel for those making the greatest contributions is admiration and a desire to work for developing our own minds' full potential instead of being ruled by our emotions. We must, above all emotional concerns, develop a taste for creativity and seek to use it whenever possible in place of violence. Civility brings society several distinct advantages. We should consider taking it to the practical extreme whenever and wherever possible. Violence should only be used to establish justice, safety, or freedom, and even in these cases it must always be seen as a highly regrettable alternative to creative solutions.

Life expectancies are rising along with literacy, education, and technology. Consider that literacy was once universally considered an extreme rarity, whereas now in many areas of the world, it is illiteracy that is the extreme rarity. Our capabilities in regard to our environment are improving. We are more prepared to deal with natural disasters now than ever before. Consider the speed with which we are able to send relief efforts to any corner of the globe. Humans have inhabited the earth for thousands of years. Only one hundred years ago, we wouldn't have even been aware of a distantly occurring disaster in the time we can now, not only find out about it, but also reach the affected area with supplies and equipment. This should shock us more than it does. We are all living on the threshold of a new age, where any conjecture as to our ultimate potential cannot be justifiably taken lightly. At what point in history was there ever an accepted code of laws and ethics that exhibit modern standards? Five hundred years ago, submarines and vaccinations were considered science fiction. Two hundred years ago, airplanes and organ transplants would have been seen likewise.

For perspective, consider one, albeit exceptionally long, human lifetime of one hundred years starting in 1895 and ending in 1995. This person would have lived through

more dramatic events and changes than anyone else in
history. The last major campaigns of the American Indi-
an Wars were fought in the late eighteen hundreds. This
person could have grown up in that environment (given
they were born in the American West) and near the end of
their life gone online to check the weather conditions on
the other side of the planet. Noteworthy events during this
age would have included the Spanish-American War, the
Philippine-American War, both World Wars, Korea, Viet-
nam, Grenada, the Cold War, the Cuban Missile Crisis, Operation Desert Storm, the moon landing, the rise and fall
of the Soviet Union, the Great Depression, the influenza
pandemic of 1918, Einstein's theory of special relativity, the
foundations of quantum mechanics, the invention of the
radio, airplane, helicopter, television, air-conditioning, sili-
con chip, modern computer, and the Internet, just to name
a few. Several steps forward and a few tremendous steps
backward were taken throughout this relatively short portion of human history. Yet we progressed despite the monumental killing, destruction, and suffering of these times.
Compare this span's steps forward to the progress of any
other century while considering that World War II alone is
generally considered to be the bloodiest and most destructive conflict in recorded history. How we've managed to
advance through the natural and man-made calamities of
time is due to remarkable human efforts and our instinct
for improving our lives.

There will, in my opinion, never be a perfect society.
The problems society faces, such as crime, will always exist to some (ideally imperceptible) degree, but when considering the state of our ancestors' world compared to the
modern world, is a society of near-perfect structure such
a farfetched idea? When considering what we have accom-

plished thus far, is there any reason to dismiss the possibility of building a better world? Can we not purposefully develop an ever-improving methodology for dealing with our current and inevitable problems?

If we do not believe we have the power and the predisposition to build a better world, what brought us this far? Was it an accident that we've built this exponentially better world from the one of our ancestors where survival was the universal, life-consuming concern? Without this built-in desire for betterment, we would be doomed to a permanent animalistic state. So long as we can build a better world, nothing else should be of consideration. It should be our highest priority. Why consider giving up on humanity when, at every turn, we see the evidence of humanity's progress? A world of unity among all nations is ours for the taking. It is a world filled with lives spent in fulfillment rather than suffering and waste. A universe of knowledge and awe is available rather than one filled with misguided ideologies and fear. The only thing that has ever moved us closer to this inevitable end is working and striving for a better tomorrow. The only thing that has significantly boosted its speed is working collectively instead of against one another. Furthermore, it is an inevitable end because there is joy in improving our lives. Progress is winning the fight. Does it sound naive? Then what are we working for, if not a better world? Are all the accomplishments of the past accidents? If so, then how is it that we've managed to make a habit of them? There is a prevalent order and established purpose to our progression from primitive times. What are the aspects of the better world you can imagine? What is necessary for it to manifest and sustain?

Form Drag

Pessimism is said to always have a positive outcome since, when the negative happens, you've expected it, and when the positive happens, it comes as a welcome surprise. This saying, in its many variations, is a feeble justification to give up on life. It is a step in the direction of taking delight in the misery and suffering of others to see one's point of view being proven correct. They can always say of the negative, "See? The world really is falling apart. There is the proof of inevitable suffering." Optimism may indeed see only the expected outcomes and the disappointments, but through these disappointments are not only the means of preventing them, rectifying them, and rebuilding, but also the hope for a better future. What other solace is there in life? The only things to be gained from the aforementioned pessimistic assumption are excuses, excuses to give up on humanity and a complacent acceptance of an inevitable doom.

There are those in the world who resist all change, and by that I mean any change. Their eyes are trained to spot it as if it were deadly. These types are often familiar with the saying "If it's not broke, don't fix it." There is a lesson to be learned from this saying. Unfortunately, some people use it as a mantra to help support a shiftless, myopic mind that misses out on the wonderful, challenging aspects of the world. Life's questions would not be difficult to answer in a simple world. Everything would line up and be easily predicted. Happily, this is not the case. The world, both natural and social, is an incredibly complex place. You can start studying any small aspect of it and still be learning about your field after a lifetime of study.

How could humans have developed anything technological ideas have produced, from the wheelbarrow to the modern computer, or that philosophical ideas have produced, from basic legal codes to the abolitionism of slavery, without facing the complicated realities of the world and, through the use of ingenuity, understanding, and wisdom, taken the difficult steps forward? How can anything ever improve without change? If change isn't allowed, then progress of any kind cannot be made. However, if society ceased changing, it would not mean that nature and the environment would follow suit. Disasters and setbacks would still occur. New viruses would continue to appear. Famine would only worsen without efforts against it.

Erosion may be a fact of life, but it can be minimized. Stopping all human change would lead to an uninhibited erosion of society. Being adaptable is one of the greatest and most useful skills a person can possess. Any who dare to remain still in this world of constant motion must be prepared to be left behind. Simple answers in regard to the world and human life do not exist, and any who think they do are wishing for a simpler world and a simpler life, which is something no one should ever covet. It is wishful thinking that comes across as a disgrace in this wonderfully complex environment of ours. There is too much to be done in life to waste it on purposefully attempting to do as little thinking as possible. There is too much to discover to waste time. This selective degradation of one's own mind is an affront to our potential.

Now many would say, "Why make this argument? You don't really think anyone wants the world to remain unchanged or to devolve, do you? This sounds too ridiculous to even point out. It's obvious that we can't stop all change and shouldn't seek to worsen our lives." Unfortunately I do feel the point needs to be made, because in this world, there are those who call technologies a thing of evil. Many of these are the same people who, given the chance, would enslave and kill anyone who disagrees with their views, and would do so in the name of righteousness. They would do so even if the very people and technologies they seek to destroy are the only things capable of curing them from a deadly condition. They would still see it all destroyed to sate their fanatical obsession with false moralities, traditions, and the ever-crumbling (and shrinking) simplistic answers they claim solve everything. They are pathologically resistant to change, and these destructive attitudes can be contagious. Those who are not appreciative of technology as a whole are not worthy of any fruits it produces, not even the added benefits of pulleys and levers. The technophobes of the world would send us all back to the Stone Age if possible, erasing centuries of progress, and for what? Is it because there, in that world, in that primitive age and no other, they might have a chance to be considered valuable to society? There they might, with their methods and outlooks, be capable of making some sort of significant positive contribution.

Why should anyone want to live in the past, in times where people were burned alive for the crime of being different, minorities were enslaved and annihilated, civility was widely defined by murderous standards, diseases, hunger, and hardships ran rampant, life spans were less than half of what they are now, and people spent so much time trying to stay alive that they had precious little spare time to pursue any other activities beyond survival? Life was hard, cruel, and terribly short. Only the strongest, healthiest, and smartest could endure in such an environment. Infant mortality in the United States today is somewhere in the range of ten deaths for every thousand births. In the eighteen hundreds to early nineteen hundreds, the number was far greater than one hundred for every thousand. So one out of a hundred die now, compared to ten out of a hundred only two centuries ago, and some want to go back to the good old pioneer days?

Every time I hear someone say we should purposefully turn back the clock, the following comes to mind: *Then do it and lead by example. Sell everything you own, and go buy some land in the middle of nowhere. There are still vast tracts of land in this country where you can go pretend you're a pioneer. Move there and, using no vehicles or electricity, build a house using timber planks, hand tools, pitch, rocks, et cetera. Don't use anything modern like insulation, plumbing, Sheetrock, plywood, or power tools, and no central heating or appliances are allowed. Then, after you've built the house, get an antique gun and some seeds. Till the soil, sow crops, and live off of the land like they did. A person could attempt and even accomplish all of this. But, when a drought, flood, or a particularly bad winter comes along and ruins your harvest, don't run back to civilization for canned goods. No, if you really want that kind of life in all of its gritty, extreme detail, then you get to do some extra hunting, skinning, and preserving of meat. Otherwise, you would do a lot more starving and dying in these more primitive times you seek to experience.*

Some don't really want what they think they want. I would like to see those times clearly myself; history is enthralling and filled with useful information and perspectives, but I also value the progress our ancestors made and the advancements of the modern world. If you are searching for a challenge, one exists in your current age. Making the world a better place is no easy task, though it is more fulfilling than purposefully devolving your life.

Others would say, "I just wish people were moral, like when I was a child." Is bigotry and intolerance on the rise, or are they fading? Is concern for the needs of the helpless disappearing, or is it being increasingly addressed by the new generations? Are bullies becoming more numerous, or are they, too, beginning to phase out due to the increasingly popular sentiment that bullies are insecure cowards in disguise? Is intelligence or physical strength more important for survival in today's world than in centuries past?

We live in an age of information. More so than ever before, we see and hear about immoral behaviors, but the same can be said about moral behaviors. It is not that people are gradually worsening. People are not degenerating into the monsters some foretell is our destiny. Knowledge naturally breeds civility, and civility allows progress to flourish. For great minds to work their wonders, the right kind of environment is needed. We must work together to provide this environment. The power of technology is simply informing us as to the true condition of our world, and though some wring their hands and say the messages are bad omens, the sky is not falling. In the distant past, it would have been virtually impossible to send a message to the opposite side of the planet. Today it can be accomplished in seconds. Is it any wonder, then, that we hear about both the good and bad aspects of humanity more so than ever before? Some people hear only what they want to hear. They cherry-pick the parts that suit their ideologies and point only to the negative to prove the world is falling apart so that their own lack of effort and morbid disillusionment will be justified. Meanwhile, the earth continues to rotate and we all continue to age. You are not worse off for knowing the facts of reality. It is an opportunity, not a burden, to use your mind, confront them, and form your own opinions—think them through, deal with new issues, embrace or denounce behaviors, conform to standards or oppose them. Information gives us the chance to learn more about the world and ourselves. Take advantage of bad news and help to prevent its repetition instead of idly declaring doom for the future. If you are not part of the solutions, you are part of the problems. One person dealing with reality, even in the most private of ways, helps to solve the problems society faces. Complaining, blaming, jumping to conclusions, and fostering a lazy mind can only add to them. Nonviolently venting your frustrations through a productive or creative outlet can be useful. Compulsive moping is always useless.

If you look for the negatives and focus on them exclusively, you will find them everywhere and will, yourself, become a part of the problem. Enough, though, on the pessimist's philosophy, the only reason to dwell on it at all is to provide a clear, easily identifiable, and shining example of what not to do with one's time or allow one's character to become.

Utopia Defined

"the form of government which communicates ease, comfort, security, or, in one word, happiness, to the greatest number of persons, and in the greatest degree, is the best."—John Adams

First we must define what Utopia is before we can effectively speed its arrival. The idea was famously laid out by Sir Thomas More in his book of the same name. There is some conjecture as to whether or not he actually meant for his work to be taken as a blueprint for an ideal society. Nevertheless it has come to mean just that, the personal conception of an ideal society. Several other classical and modern authors have written on the subject, laying out their own versions of what the ideal society is and how to reach it. From this, others took the opposite direction and wrote about the nightmarish futures the world could potentially become. Some serve as warnings, such as George Orwell's infamous book, *Nineteen Eighty-Four*, which details the inherent problems of excessive authoritarian control. These nightmare worlds are called dystopias. All combined, these fictional and/or philosophical works on utopias, and anti-utopias, number well over a thousand titles.

Some think we should not strive for such currently nonexistent ideals as we can imagine. They say we should only focus on the small changes we can make in the short run, and let tomorrow take care of itself. It is odd, however, that living your life by this same concept would be generally discouraged by many of these same people. For instance, it is doubtful that a caring parent would tell their child to make no plans for the future, due to a perceived limitation of human potential. Why not establish distant objectives? Why not take the action toward positive goals regardless of their difficulty? Why not look to those more ultimate goals, which, granted, we may never see to fruition, but that future generations may solve? The approach of condemning the second step in the name of the first step's immediacy should not become habit. Due consideration should be paid to the facts of today and the immediate need, while simultaneously monitoring and considering the long-range objectives. Acknowledge obstacles for the daunting difficulties they are, and work to overcome them. To attack utopianism for its foresight is to denounce strategy for concern of tactics. This is a weakness, as both are required to form the most efficient plans. The trend has been to only progress when necessary, as in, when collapse looms threateningly. It would be of use to change such habits. When required, we can make quite an impact. If society took the attitude that betterment is a "have-to" issue, then we could gain ground faster, and history would judge our generations with greater favor. It would judge us all as builders rather than as complacent bickerers. There is much work to be done. The third assertion is that since we are capable of building a better world, it would be a shame to do otherwise.

Adolf Hitler is quoted in his political testament as saying, "The universalists, the idealists, the Utopians all aim too high...I myself have always kept my eye fixed on a paradise which, in the nature of things, lies well within our reach." Being (hopefully of course) a realist, I would say that those who place limits on human potential are historically prone to being proven wrong, and focusing solely on immediate goals is for the shortsighted. However, we must be ever careful of the nature of our plans since one man's paradise can be another man's hell. The ideal state is one of maximum advantage for all civilization. It considers and addresses all of humanity's concerns, not sole, vaguely defined cross sections of the world's population.

Disaster Control

Why pursue a Utopia? One reason is that survival is worth pursuing. Science declares that all life on this planet will eventually be extinguished. There are numerous theories as to how human life may continue after the point in time is reached when life on Earth will have met its natural end. In the meantime, super disasters potentially threaten our survival. Environmental and medical threats kill us and destroy our works. This is not to mention the threat we pose to ourselves. Is it sensible to predict (say for the sake of an example) that a super-volcano is going to decimate a country or the entire world one day and respond to this prediction by taking little or no constructive action to prevent the disaster? How ineffective would the governments of the world appear in retrospect? With few humans left alive, would these survivors be forced to remember that, prior to the cataclysmic disaster, their governments focused attention on comparatively useless issues? What if they possessed the combined resources to prevent some hypothetical disaster, but failed to act? Would this not be criminal? Would not the destruction of an entire nation or continent through neglect have to be considered a crime against humanity if anything can be?

United, our response to disaster is stronger than if we are divided. Lives and resources will be saved to a greater degree in a state of cooperation. Disasters that could not have even been foreseen in the past will be averted in the future with the aid of enhanced technology and knowledge. Reaching Utopia is our best chance for survival regardless of what the future may bring. To say we can't achieve these ends because of stubborn attitudes makes us look like feeble incompetents. Is it to say, we could master the awesome power of the natural environment if only we could agree that unity is worth pursuing? By preparing for super-disasters, we learn about their smaller versions, how to minimize their damage, prepare for them, rebuild, and ideally avoid them altogether. This applies to all potential threats to our survival, and all should be addressed. These efforts would yield extremely beneficial side effects.

Consider the bubonic plague that once decimated Europe. This disease still exists today, but medical science, working alongside societal improvements, has minimized the problem so that it is no longer the threat it once was. We possessed the potential to deal with this epidemic when it devastated nations in the past, but lacked the knowledge and technology to control the outbreak. The people who died from this affliction died needlessly. Their lives could have been saved if humanity had only progressed at a faster rate, instead of spending all that time and effort trying to kill one another over the ideologies and subjective aesthetics that have now been proven worthless through retrospect. This is not to condemn our ancestor's lack of progress, since there is good reason to celebrate the abundance thereof, but the attitudes and achievements we expect from others should always be thoroughly laced with an understanding for circumstances, critical contemplation, and perhaps most importantly, piercing self-reflection. It would be to our advantage to learn from the disasters of the past and try to prevent further catastrophes as much as possible.

These efforts have always wholly benefited humanity and often in unforeseen ways.

The Debatable Conceptions

One of our problems in accurately defining Utopia is that there are so many different views of a better world that they conflict. Your ideal society probably does not coincide with mine. If common ground is sought, within your own community, you will find opinions of such variance that often points of severe contrition will arise. This is not to mention that in foreign cultures, people are raised with exposure to very different ideals from our own. Entire nations of people are taught from birth to value a different governmental structure than the one we've come to more or less agree with in the United States. If there is an ideal global structure that can work for the entire world, how would we ever arrive at it with such diversity of thought? Many have pictured the ideal society as having no government at all. Many have pictured it as having more borders and laws rather than less. Some say a state ruled by a universal belief system, like a religion, is the key, while others claim the leading factor of an ideal society is found purely in its political system. Agreement is evidently, in the sea of propositions, hard to come by.

Let's see if we can find some common ground. Granted, we may not agree on specific goals or methods, but hopefully we can agree on the basic points. The obvious necessities for building a better world should be much easier to identify than its finer points.

Taking the aforementioned factors that define a successful society (assuming you agree on the criteria), we can begin to move toward the goal of defining Utopia more clearly. Let's review and take them each in turn. Again, the factors are civility, safety, prosperity, knowledge, health, technology, independence, comfort, efficiency in dealing with societal and political issues, and the preoccupations of the collective will.

Note that to fit the criteria of a social imperative, the factor must not be something that would be ideally left almost entirely to the government. If it were, then it would be a governmental imperative, like justice. All of these aspects are, in part, if not entirely, a responsibility of society.

Civility

Unity is beneficial. Everyone should be able to agree on this statement, at the very least in this general form. Most corporations both preach and illustrate its validity. Families put it to use instinctively. Society, as we know it, could not have been built without cooperation. We mustn't take its benefits for granted. When people work together, they accomplish more than when they work against one another. Why, then, is a one-world government so widely condemned? Why do we habitually look at the world through the lens of fear? So many of us want to be ready in case the world begins to fall apart, but how many of us say we want to be ready in case world peace is achieved? Many wouldn't even know how to face such a situation. It is inefficient and irrational to only consider a single outcome when an infinite number of possibilities surround us.

The degree to which we are united has historically been largely dictated to us by our circumstances. When cooperation is mandated, even sworn enemies can work together to achieve a common objective. When little can be gained from cooperation in the immediately foreseeable future, we have enslaved one another. This is called being ruled by your environment, as opposed to mastering it.

What nation has directly benefited from closing its borders and cutting itself off from the world? When has working together peaceably toward mutually beneficial ends directly produced detrimental results? To deny the possibility for successful global unification is to accept the everlasting threats of discord, and war, as inherent inevitabilities. Some can live in civil harmony. Therefore, we possess this potential as a species. It is beneficial. Therefore we should seek it. People might say that this kind of global cohesion would lead to a dystopia where minority classes are subjugated by the majority since much harm could be done in the name of the greater good. It should be taken into account that this is already a valid concern. There is no harm in unifying our efforts, especially if those efforts are to build a better world for everyone. To say that the mass sum of our combined efforts must always yield negative results ignores a great deal of the positive evidence. To say global structure could make it too easy to bring about negative results is the same thing as admitting global structure would make it easier to bring about positive results. Either way you are acknowledging unity's potential. These concerns are entirely valid, yet they are not very imaginative or constructive arguments since they are simply dismissive. The largest empires of the past failed because their systems were far from ideal and because they lacked the capability to establish far-reaching stability. In modern times, the potential for stability has grown globally, along with civility, due to our ever-increasing capabilities.

Unification (on any level) under positive principles yields greater benefit than does division under a mixture of principles. For the principles to be wholly positive, they must give room for the ideal degree of self-rule while encouraging civility and structure to the highest degree. If civility and basic freedom cannot be agreed upon, then granted, a one-world government could not function. These values are sure to win the popularity contest over time, though, due simply to their demonstrable worth. The key is that, in time, the trend will prove people the world over do value their responsibilities even if they, from time to time, side against them out of laziness or fear.

There can be a global rule that does not ignore local representation. There can be a global rule that gives everyone everywhere equal rights and pays due attention to macro- as well as microeconomic situations. This system already exists in some form since there is a current degree of global social, political, and economic cooperation. It just isn't as developed as it could be. Problems could be addressed more efficiently if we were all on the same team, especially the big problems. If an ideal system (social or political) can operate over a span of three hundred thousand square miles, then with the desire for unity, it can operate just as well across over three hundred million square miles. Over time, the argument has popularly shifted away from one of geographical logistic concerns (since we can now master them to such a great degree) to one of cultural agreement. What works for us won't work for them. What works for them won't work for us. These are feeble excuses. The ideal system works well for everyone. That is what makes it ideal. So one might say this kind of ideal system doesn't exist. Do people who don't believe in having specified rights turn into manic degenerates when they are given independence, or do they strive on? Do people who lose rights cease to make any of their own decisions, or do they continue the struggle for prosperity? To say that they can't function in our system or that we can't function in theirs is to say that successful immigration is an impossibility. And yet, many people do it for immediate personal gain, and they tolerate whatever system they join in exchange for whatever they seek (better jobs, travel, economic opportunities, love, etc.). The ideal system is not based around morality, tradition, or any of the things we would see as barriers between us and the people to whom we most differ in the world. The ideal system is based on providing specific services logically, effectively, and efficiently. The government must have a clear set of exclusive objectives in order to achieve them. Following the drive for betterment, a system can be built that yields the greatest return for effort spent, just as with any other endeavor. We should not seek to conquer the world through force and to build a one-world government. We should seek respectively to define, implement, and promote the ideal system with civility and tact. This ideal system will then sell itself and end up conquering the world due to its own merit. Countries have always emulated methods that prove to be of value.

Many would say on this point that unity, while being a good thing, should not be looked at in such extreme terms. What then is our ideal condition? Do we remain separated for all time due to tradition or petty inconsistencies no greater than those we already have with our fellow citizens in a free society? What is our ultimate goal? If we aim for mediocrity, then the true danger is in achieving it. Should we not aim as high as possible? Aiming for easily attainable goals and focusing on the smallest possible steps has undoubtedly led us gradually forward, but at a much slower pace than we are capable. Large steps are feared, in part, due to the valid concerns of caution, but time is also of the essence. Children are dying while we debate over trivialities. Once in the realm of near total cooperation, with it in attainable sight, it will be gained simply to make life more efficient. Seeking global unity should be our aim if for no other reason than that it is a positive social goal. To dismiss cooperation as unattainable is naive and weak. Some people can unite, but others cannot? If everyone simply valued this goal, then the point would be moot anyway since we would already have it. People must see its unquestionably positive value and desire its benefits before we can achieve a greater degree of global cooperation. It doesn't have to happen overnight, but cooperation should be a high priority nonetheless.

Cultures should be preserved to continue yielding the rich diversity that facilitates the multifaceted ideologies we've created with our various unique perspectives. There is advantage in cultural exploration so long as people are free to choose their own individuality. However, when a culture becomes a personal mandate rather than a societal, goal it becomes a tyranny. Local representation, with the constraints of well-formed, specified personal rights and responsibilities, can remedy any rifts between struggling multicultural societies. Limits can be placed on multicultural goals in the interest of preserving the native culture, without resorting to oppression and injustice. The more society can accomplish without the use of governmental force, the better.

Wouldn't we all benefit from world citizenship as opposed to any particular national citizenship? Wouldn't we all benefit from a universally recognized and enforced set of legally sanctioned rights? If not, then what have international courts and the United Nations attempted to establish when the world has been called on to provide a universal form of justice and peace? Why do we continue to only act on such matters when absolutely necessary? What gives any country the right to punish tyrants, excepting that if middle ground can be found among the justice systems involved in the trial? If middle ground can be found, why is it not further developed and pursued as a group objective? Why only develop this evidently highest level of authority when absolutely necessary or convenient? Theoretical situations should be presented when there are no current cases, with the aim of testing the system. Law is not supposed to be enforced only when it is convenient or necessary to do so. National law is complex, and international law becomes maddeningly complex when trying to consider all systems, procedures, and their various levels of administration to reach a middle ground where justice is effectively administered. However, this complexity and variance of opinion is not a valid excuse for ignoring this obviously advantageous objective. International law should be developed, and any nations united in civility should be expected to enforce certain basic standards. Those administrations that ignore these basics should be shunned, in order to encourage participation in global trade and civility. These efforts of communicating and forming agreement between nations have the potential to create grand-scale cooperation the likes of which human history has never known. It is in this seldom-tread field of what can be universally agreed upon as justice that lays the fertile soil of Utopia's justice system, whatever it may be.

In many societies the movement of people from one place to another is seen as an invasion of sovereign territory, whereas in Utopia this concept is alien, since there are no enforced borders. The only maps that show them are called historical maps. In our world, culture is seen as something worth killing over; in Utopia it is an aesthetic that links us to the past, and everyone lays equal claim to it. In our world, wars are fought over material and political gain. An act of war can be something as slight as the slander of an ideology, or a disputed border, drawn and redrawn in centuries past. In Utopia, death is seen as an act of war against life, and everyone is willing to fight this war against whatever caused the death.

What are borders? Try explaining their function to an imaginary outsider of our world or to a child. Borders are monuments to uncivilized historic events. To say that a border is strong is to say that it is well defined. Evidently the nations involved have an interest in keeping the line drawn. Never are they stronger than in times of war. If unity is an objective, borders are obstacles. Lines drawn in dirt are hindering our potential. From a distant perspective, they appear as childish relics of a time when division was seen as, at best, a necessary evil, at worst, as inevitable. We have too much work ahead to continue wasting time on trivialities and overly strong borders are not helping the situation. Borders are not helping the overall situation. I'm not suggesting we tear them all down with immediate force. I'm suggesting we develop a methodology that causes them to gradually disappear from the maps on their own. If two countries (not necessarily neighbors) were to agree to a great extent on their aims, methods, and structure, they could merge with little difficulty. If everyone were focused on the same goals, borders would disappear effortlessly. As things stand now, trying to actively make some borders disappear would be like trying to move the moon with a thread. However, as cohesion increases, people will eventually see even the strongest of current borders as nothing more than traditional formalities. When this happens, their dissolutions are imminent, and unified efforts, instead of being seen as threats, will be seen as obviously rational responses to the need for efficiency. As Utopia approaches, borders will crumble into historic features, cultural ties, and geographic mental aids.

Safety

Safety from harm should also be easy enough to agree upon. Indeed, even many who see national divisions as inevitabilities would say that the successful society is free from turmoil. Only a mindless thug or a bloodthirsty lunatic believes violence is a goal. It should be avoided, an option of last resort at best, used only when no other viable option will work.

There is no perfect protection from danger. In all forms of government, the people are expected to defend themselves from danger to some degree. In Utopia, this thankfully inescapable element of self-preservation is understood, monitored, and compensated for by the system, though the goal is to minimize the need to defend oneself by creating as civilized an environment as possible in the first place. It is one of the basic functions of any government to provide reasonable protection from harm. There will never be a perfect police force or a perfect judicial system. For the sake of realism when envisioning the ideal society we should expect that crime will always exist to some degree (though it may be useful to speculate on what would need to be done differently if it were marginalized or perhaps even eradicated entirely). The threat of violence will more than likely never be completely removed by the advancements of society. Given this is the case, we will always be in need of police departments. Even in a much more peaceful environment they can perform functions which would be difficult for society to establish on its own. Businesses often see a need for greater security and so they hire private security workers. In the areas of the least amounts of crime they are not needed. So as prosperity increases, the tendency to commit crime decreases. As the tendency to commit crime decreases, prosperity increases. The ideal police forces' functions must be clear and its contributions to society obvious and prized. There must be no 'optional' laws. Laws in a Utopia could not be allowed to become the mockeries that too many are today. For the laws to be effective they must be well known among the citizenry and strictly enforced. By definition, Utopia exhibits lower crime rates, making law enforcement a less dangerous job than it is in many parts of the world today, leaving police free to pursue the additional, less crucial services they may provide to communities. Greater attention to activities such as safety education, searching out opportunities for accident prevention, as well as being the positive role models and helpers they are meant to be would yield a substantial boost to society's betterment. Crime, misallocation and mismanagement of their efforts, lack of proper controls, and injustices all work to minimize anything beyond their basic functions and even serve to disrupt and complicate these efforts as well. For their contributions to be maximized, they must be encouraged through reward and enabled through progress.

National security, on the other hand, would be best achieved through unification. This would reduce the threat of war more than any other action possibly could. Along with this practical elimination of war, there would also follow a lack of need for weapons of mass destruction. What good is a neutron bomb or a chemical weapon in a world of unity? There would be no legitimate need for them. The only theoretical need for nuclear weapons would be if we needed to destroy some threat like a giant meteor. Imagine if we could shift half of our military resources to natural threats and the other half to civil threats, instead of spending billions on international threats. This would yield a much safer world for all. The threat of attacks committed by individuals against society cannot ultimately be as suppressed as the threat of war. These terrorist-type attacks can be more effectively minimized through a stronger police presence in the areas of greatest vulnerability. For instance, every school should have a police presence, and with any available spare time, this presence can serve additional positive functions in a vast array of ways, such as breaking up playground fights, providing a positive influence, and discouraging behaviors that may lead to future criminal behavior if left unaddressed. Perhaps even school life could become more civilized. As individuals, we too can aid in the creation of a safer world by watching for potential hazards and reporting them or dealing with them responsibly. The safety of human life and our environment should be a high priority.

Prosperity

We can practically skip over the justification for valuing prosperity since, if absolutely nothing else, everyone's picture of Utopia is one of a prosperous society. Few, if any, of the other aspects of Utopia can be obtained without prosperity since, without it, the society in question is up for grabs and sold to the highest bidder. Prosperity is a combination of economic success and a strong general welfare for all members of society. It is an exceptionally easy aspect to envision. The far more difficult question is how to best achieve it. How much responsibility should fall on society and how much on the government in this quest? Just as safety's concerns are a combination of effort from both parties, so prosperity can also be seen as a team effort. It is enough to say for now that this factor should not be measured as an abundance of wealth. It should be measured instead as a lack of poverty. The economic goals should seek to eliminate poverty as a primary consideration. There are many methods that propose the ideal in this regard. To accomplish the monumentally difficult problem, study the nature of poverty and find the method or methods that have best eliminated it in the past and then attempt to put them into practice. Take the challenge a step at a time, progressing toward maximizing the poor's chances for betterment while preserving the rights of all people to succeed in life. Consider all worthy considerations. We must fight the economic struggle as if it were a war if we are to see any progress within our lifetimes. Any progress made ultimately speeds Utopia's arrival.

Knowledge

The value for knowledge can be found in even the most primitive societies. Elders teach the young how to survive. They hand down their own values and beliefs. The young are expected to learn in order to become the people their elders see as valuing society and to aid them in leading fulfilling lives. In Utopia the love of knowledge is taken to its utmost extreme. Children are instilled with a hunger for learning at a young age. Everyone wants to discover as much as possible about the world while they are alive to experience it. Museums, parks, wildlife preserves, and libraries are among Utopia's most cherished creations. Utopia's citizens have a greater access to these resources since they are more widespread and developed. Here, reality is better appreciated. We are, after all, organisms living on a giant sphere with water on it that orbits a massive ball of nuclear fire, and all of this inside a vacuum of unfathomable size. There are microscopic cells inside our bloodstreams that exist only to fight and die to keep us alive. Without them we would not survive for a single day's time, much less build machines to monitor our heart rates, among the millions of other incredible things we do that are nearly unbelievable when viewed in certain lights. Nothing can be truly appreciated until it is well understood, and there isn't enough time to fully understand any subject we encounter. Everything is in constant change and can be subjected to never-ending scrutiny. How wonderful it is that life should be so overwhelmingly intricate and involved. It is as deep as you want it to be by simply changing your perspective. You could never learn enough to become bored by it. There is always another more difficult challenge just ahead.

Our primitive ancestors looked to the night sky and saw the stars as what must have appeared as pure chaos. Nightly their patterns changed. One night's careful observation would have brought the idea that they are moving around in the sky. A month later the view would have changed so drastically that the original map they set out to draw would have changed drastically. At this time they could not have grasped that they stood on a rotating sphere, or that it was in motion through space. They could not have understood that the sun is one of these stars, only in closer proximity than the others. They had no way of knowing that some of these stars were not stars, but distant planets. Would they have believed that there were vast numbers of stars just out of sight that could be seen if they were to look through something called shaped glass?

We take these elementary concepts for granted. The most amazing thing of all is that we are alive, and here, to witness these wonders and to be able to comprehend them. The point is, our primitive ancestors observed a phenomenon and, trying to gain answers to its nature, used several methods to explain them away. The path that has led to a greater understanding of our environment is the path of painstaking study and documentation. They catalogued, charted, and mapped the sky night after night to build the foundations of modern astronomy. The foundation they created has led to our current knowledge of still more difficult to understand features such as black holes, nebulas, and quasars. This difficult path of thorough documentation and study were the first bricks of the modern library of knowledge. We should pursue knowledge of social science by similar methods, aiming for betterment through the discovery and development of the positive features of humanity.

Throughout our history we have increased both our total available knowledge and the knowledge of the average person. As it grows, where else can it lead but to the vast improvement of our lives? Can you visualize its next big step?

Health

Utopians possess greater health than we do. It is another aspect in envisioning them that exemplifies their society. With an appreciation for life comes the desire to get the most out of it. This means prolonging it to the greatest possible extent. The masses see health as an attainment worthy of pursuit. As well as day-to-day health concerns, their medical knowledge and technology being greater than our own enables them to lead healthier lives. Many afflictions that kill us are, in their world, easily treatable. How many afflictions can we cure? How far can we extend the human life span? How can we best improve the overall health of the masses? How can suffering be minimized? These are the questions medical experts face. In Utopia these questions remain. Methods are consistently improved. Lives are saved. How far can we take our medical capabilities? Will our life spans always be roughly the length they are now? How far will we be able to usefully stretch a human lifetime? Will nanotechnologies break current barriers? Will mechanical implants become common implements? There is much speculation, but that is all it can be until progress is made and theories are tested. Progress is being made daily, and it will continue. Looking ahead, Utopia's health system may seem like science fiction, but whatever else it may be, it will be more efficient and effective than our own.

Technology

Regarding technology, can you imagine a world where it has provided greater capabilities than our own? Imagine technology stopping a forest fire that would now rage out of control. Imagine being able to reach the survivors of a transportation accident in half the time it would currently take. Nonmedical technologies can save lives as well as the medical field's efforts. Imagine a world where machines help to create an environment where less effort is required to perform tasks. One could say we already live in this world. Potential improvements are waiting to be made. Technology can enhance all of the aspects of an ideal society. Since it can aid our knowledge, it can help us build a better world. It can help to protect us, and make our lives more comfortable. It can help to improve the economy. A person can communicate with people on the other side of the world in several different cultures, all due to technology. This capability encourages the capacity for unity and understanding, whether we employ it or not is up to us. Technology is our greatest physical and mental aid in achieving our potential. From the earliest implements to the cutting edge, it is continually growing. If only we could advance at such a rate socially.

Independence

Just as borders are monuments to conflict and separation, the Pyramids are monuments to slavery and oppression. They are monuments to thousands working their lives away to build a tombstone for one person, a person so self-obsessed as to believe they could carry their possessions (including their slaves) with them into the next world. This type of person would be seen today (and rightly so) as a lunatic. No person or institution has ever been worthy of recognition by force. Imagine what could have been done if the pharaohs had instructed their people to live free and build a better world for themselves rather than gigantic tombs. The Pyramids should be preserved, if at all, for the same reason Auschwitz should be preserved, as historic relics and reminders of mankind's obvious wrong turns.

What is the ideal degree of self-rule? Some would say that total self-rule is the ideal situation. Trying to imagine this form of theoretical paradise, I can still foresee situations where a government would be useful. The unforeseen is inevitable. The organizational power of a group seeking proper governmental objectives will, in my opinion, always be of use to society. Whenever the proper objectives are strayed from, even to the slightest degree, efficiency and effectiveness suffer, often imperceptibly, due to the limited scope of viable comparisons. This "proper scope of governmental operation" is a vague, highly debatable concept. We've yet to discover what brings the best results, though we all have our own opinions.

Those who do not value freedom are really searching for someone to blame when events turn out less than perfect in their lives. What are the historic results of turning your moral conscience over to someone else? Those who are eager to surrender their will have no conception of creativity, much less a value for it. If an abandonment of all responsibility were to become the majority's overriding desire, then the situation would become completely hopeless since no elected official, once given the job of guiding society, would know what to instruct beyond some form of instinct, poor material for developing long-term plans. Again, the good news is that we are moving away from the widespread slave labor conditions of our distant ancestors. A love for life and choice still grows. Contempt for the lazy mind that finds no value in freedom is a virtue that has helped to secure the freedoms we currently enjoy. The desire for greater responsibility marks the progression toward Utopia and away from the dire assumption that people must be forced into uniformity to gain the most good.

Society's needs created government, and these needs maintain its existence still. The desire to reach Utopia was at work in the ancient times that spawned the first form of structured civic order, even if the individuals involved didn't realize it at the time. They only saw the immediate gains of organization and a localized unity. They didn't realize this was a small step toward a world they could not even imagine at the time. Never in their wildest dreams would they have considered that, by working together, the wonders of the modern world could be achieved. Structure exists because there is benefit in it. Government is a direct result of the instinct for betterment. There is value in the properly functioning government. We should not seek its elimination (even with good intentions of self-rule in mind). We should not seek to end anything that produces obvious, highly advantageous value. If the need for rule were to naturally dissipate one day (I don't think it ever will), then we could put an end to it without damaging society in the process, but for the time being, there are much more critical issues than to look into eliminating the greatest of our advantages to achieve a theoretical ideal state. Again, the ideal state will form itself through trial and error, and over time will rise through the proof of its value. Simply possessing power does not justify its use against societies or governments.

Comfort

Comfort could, in this context, be defined as the degree to which these other goals are realized. Prosperity, technology, knowledge, and health are most definitely required before comfort can be established. In a poorly functioning society, survival is the ultimate preoccupation. There is no time for study or discovery; there is only time for necessity, and thus many activities that could better the overall situation are missed. What good is a museum in a community rife with starvation? If the masses are too busy focusing on mere survival, there will be no time for other critical developments. Even in ancient times, general comfort was used as a way to gauge the progress of a society, and many of their philosophers saw it as a necessary prerequisite of societal betterment.

To provide an environment where comfort can be achieved, we must first possess a stable, effective economy. Without some degree of prosperity and some degree of peace, there can be no real comfort. The value of comfort is in giving people the ability to pursue their personal objectives to a higher degree. A person with spare time can build a better device or invent something of benefit. Through relaxed daydreaming they may envision some great enhancement to a current process. By pursuing a hobby, they may revolutionize a field of study or make a groundbreaking discovery. Through study they may develop new skills. They may find they have a talent that has escaped their notice. Advancements are hindered in an environment where survival consumes every moment of the average person's time. There are too many such uncomfortable places in the world today. Potential scientists, doctors, engineers, teachers, and leaders are starving in inescapable poverty. We must work toward eliminating poverty to allow for the facilitation of comfort, though that is not enough to gain its full advantages. We must also encourage each other to be productive with our spare time. We need to use every second, since time is fleeting. Knowledge must also be achieved to utilize our spare time in a positive manner. Comfort allows for a greater exploration of our environments and ourselves. To harness and enhance our full potential, comfort must be established as a direct result of Utopia's other identifying qualities.

Efficiency in Dealing with Social Issues

Efficiency in dealing with societal and political issues is, as previously noted, a key aspect of the Utopia to come. When a disaster occurs, we typically see a greater, more effective response in this day and age than in years past. Communities pull together in hard times. They rebuild the destroyed and repair the damaged; they self-maintain to an extent. However, efficiency in dealing with social issues is not ideally measured in such terms. It would be better measured as how society handles issues that are not considered disasters. How does it respond to everyday challenges and issues that have been confronted yet remain unsolved over great lengths of time? Does it make progress or struggle to no end? The future world will have learned from the mistakes of the past. It will be aware of the faster methods, the more thorough measures, and the least costly. It is not enough to simply fix the problems that arise along the way. They must be fixed in such a way that we can be pleased with our competence. These issues may involve the difference between life and death and should be approached with all seriousness and concern. If an aspect of our society is hindering our potential, then it must be addressed. The objectives must be clear, and secondary concerns may need to be prioritized in order to achieve the most important goals. The ideal society responds effectively and efficiently to the inevitable obstacles of its progression. In short, Utopia's society values betterment and seeks it as paramount to other concerns.

The Preoccupations of the Collective Will

This is the measure of society's concern for the aforementioned parameters and the desire to refine and explore aesthetic values. The concerns of Utopia's collective will are obviously aimed at betterment. They expect progress from their leaders. They expect their political system to make improvements to its efficiencies. Stagnation is not to be tolerated in the ideal society's government. Efforts at building the better government must be a continuously explored imperative. Utopia's majority wants the aforementioned qualities of a successful society to their greatest extents, and all are preoccupied with improving them. Societal goals include (secondary to the chief imperatives) the beautification of the world, the desire to turn the world into an ever-evolving work of art. Nature must be maintained and preserved, pollution minimized, and aesthetics explored. The corporations of Utopia will spend much to create safe, desirable environments for their workers. There are unapparent benefits to these aims. The collective will strives to enhance all aspects of life. When new discoveries are made, the news is spread with unprecedented speed since everyone is deeply concerned with progression and advancement. They all want to know how they can help, where the greatest efforts are required. In working toward the goals with such cohesion and organization, everyone is aware of the problem areas. The better our understanding of these areas of difficulty, the greater are our chances to solve the toughest dilemmas. This system of universal progress monitoring may sound like fantasy, but in Utopia it is seen as a tool born out of common sense and utility. With unity comes the capability to focus our efforts. When we possess the full strength of this capability we will, with organization, begin to see our world transformed beyond any of our expectations. As the ideal state nears, the numbers of those wishing to speed it along and make their own contributions will steadily increase.

Personal Definition

As I've already said, I don't believe in a perfect world, only in an ideal methodology. What do you see when you look to the future? What does the very word "futuristic" bring to mind? What do the textbooks of the future contain? Utopia is advancement beyond the point of universal accord. It is exceeding the expectations of our capabilities. In Utopia, famine is something children read about in books rather than experience firsthand. The citizens of this world realize that from the time they are born they are writing their obituaries, and they seek to write them as well as possible. It may sound like fantasy for the human race to achieve this kind of progress, but in looking to the distant past, where does it seem like we are going? Given the opportunity, wouldn't you want to help build the most beautiful future possible, if for no other reason than to see how far our potential reaches? We can create a better world for future generations who are sure to be more advanced than ourselves. Imagine what they and their descendants will create if they are given the advantage of a strong foundation.

Moving on, all endeavors worthy of consideration aim at the Utopia where we enjoy the earlier mentioned attributes of a successful society to their fullest. If all roads worthy of consideration point to Utopia, which ones are the best to reach our destination? We must remember that at its core, Utopia is not only a place, it is a path itself. It aims at continual improvement. It looks to refine and advance society indefinitely. Perhaps you can imagine a better world than I can. The citizens of Utopia all can because they know what works and what does not. They have the benefit of hindsight. Utopia is the pinnacle of human achievement in action.

The Problems with Pride

"The race enmities and prejudices are decidedly waning."—Nikola Tesla

Pride is often a poorly understood, vaguely defined aspect of our lives. I see it as a bit of a social problem, due to, if nothing else, its controversial nature. There seems to be a debate over what it is and what to do with it. It is a facet that needs to be seen in a better light for us to move forward. Of course everyone thinks they have a good grasp on their own appropriations of pride. It's always the other person who has a problem with thinking either too much or not enough of themselves. What are its uses and misuses? Does it have any value? Many have said it should be wholly eliminated. Almost all views condemn its excesses. The following is an attempt to identify the trait's proper place in our lives.

Is pride a positive or negative trait? Its definition suggests it can be either. Arguments could be made both for and against it. Most would agree that there is an ideal amount of it for each individual, largely depending on their actions. In other words, it has its time and place. Without this feature of our personalities, the world would not be a better place, as some may contend. When it is justified, it can be useful in guiding us toward a better, more fulfilled life. When it is unjustified, it can lead to ruin and delusion. The key to pride is that only when it is justified is it beneficial. When it is unjustified, it is always destructive to the person's personality and character. It can even become an addiction.

Does it sound like I'm attacking a harmless aesthetic feature? If so, consider you may be viewing the pride from the standpoint of daily life. Looking at the big picture, pride is much more than a triviality or a simple reflection of character. Its misuse has led to some of our greatest setbacks. The glaring examples are, of course, genocide and slavery. It was used as the justification for the greatest atrocities of World War II. It has been a justification for slavery and subjugation throughout the ages on all continents: "What gives you the right?" "We are better than you." This is the prideful response.

I think pride is often the real reason as to why many murders are committed in what are often seen simply as acts of random rage. Many serial killers have been looked down on to an abnormal degree as children. This lack of self-esteem can develop into a hatred for life and the happiness of others. Several have also used a sense of narcissistic superiority (excessive pride) to justify their crimes not only to themselves, but also to the world. Too often we forget how important our day-to-day actions can be. Everyone has a breaking point to being looked down on and treated as less than human. If treated this way long enough, the person may stop acting like a human to cope with their situation. So also the effects of excessive pride can lead to a delusional obsession over one's own worth. It creates in some insane fantasies that the individual will go to detrimental extremes to maintain. Putting pride into proper perspective can be difficult. We cannot afford to take it lightly if we want to understand and identify certain social warning signs, both on the personal and national level.

How do we know when pride is justified? This is the core question, and we'll come back to it in a moment. I would first like to clarify why I mention the subject in the first place. In the Utopia of the future, there must not only be technological achievement, there must also be social achievement if we are to create the great society that is ours for the building. In order for society to deserve this Utopia, for it to be able to properly appreciate and utilize its awesome capabilities, we must gain a greater understanding of proper values. We must treasure our short time in this world and try, as a team, to move forward and constantly improve.

Take where we are now, for instance. The argument for or against a certain form of government can fill volumes and has its usefulness in finding answers. Evaluating our options is a key to making good choices. The form of government Utopia has (whichever one is ideal) must be chosen because it has proven itself to be better than the alternatives. But, there are other less helpful arguments, arguments so elementary and tired that the vast majority can see them as belonging in another century. There are many issues we would like to finally move away from, yet they continue to linger. Slavery has been greatly reduced in practice since ancient times, yet are its false justifications well known among the masses? One would think that in this day and age, racism would be a thing of the distant past. Unfortunately, this is not the case. Children are keen to imitate the base behaviors they observe. They can take to a belief like racism easily if in the wrong environment. Does it have a place in our future? Has overgeneralization ever had a shred of usefulness? On the playgrounds of Utopia, instances of demonstrated racism or bullying of any kind would be rare at best, since common knowledge will have, by then, brought such expressions into a brighter light. They would be seen as so transparent, primitive, and uncivilized that to even pretend at them would be completely unfashionable. How long did it take humans to learn as a majority that murder or theft should be considered a crime in a successful society? It took a long time for humanity to learn the simple lessons that today are so universal and easily taken for granted. More lessons have followed, and more still lie ahead for us to adopt collectively. They should be welcomed with open arms instead of dreaded as challenges too great to face.

What is racism? Many would answer that it is a hatred for a race or races of people. This is not the case. No one could argue that hatred is an obvious byproduct of racism, but that is not exactly what it is. As defined by the *Merriam-Webster Dictionary*, it is: "A belief that race is the primary determinant of human traits and capacities and that racial differences produce an inherent superiority of a particular race." Pride is the root of all forms of racism and prejudice. Even when hatred is the face of the issue, it is a mask hiding an arrogant, insecure, and illogical pride. Racism says, "My race is better than yours." There is no logical argument that can justify this arbitrary position. It is cheap emotional overgeneralization and an attempt to create an absolute that cannot be validated. If we generalized in this way continuously, we would soon find ourselves condemning the entire human race, because in a search for negativity, we would find that all peoples possess negative as well as positive traits. In defining racism, Ayn Rand put it well in an article that was included in her book, *The Virtue of Selfishness*, "…it is a quest for the unearned" Along this same train of thought one cannot help but conclude that it is not only an overly convenient "automatic self-esteem" as she calls it, but also a laziness of the mind, a refusal to consider the individual simply because it is easier to accept or condemn entire groups of people. Without these useless generalizations built on subjective statistics, which the racists claim mean everything, they would have to think. They would have to judge each person without useless prejudices.

Just as natural selection causes unused characteristics to fade while employed traits are developed in animals, the same can be observed in the human world. The useful is used, and the useless is abandoned. If a group of primitive peoples have used a certain obscure, unique tool (for survival, say) over many generations, would it come as a surprise that a person born from their lineage would, on average, have a higher natural aptitude for the use of the tool than would a foreigner? Does this mean that the foreigner would always have a lower aptitude than a native? Would there not be the occasional outsider who is able to become just as capable with the tool as the natives? Would there not be the occasional native who shows little or no skill with the tool? Are there not various anomalies, such as a savant child being born from average parents, or a healthy child being born from sickly parents? Such is life: few absolutes and few simple answers. Neither physical nor mental aptitude can be accurately predicted based on ancestry. To attempt to do so is a futile escape from reality, not to mention a complete waste of time.

Laying waste to the argument for racism is no great task, even if there are still many who believe that the world is such a simple, terrible place. If the racist ideology in question is to the extreme (as is often the case) and asserts that to bring prosperity all other races should be destroyed, the task of arguing against the idea only becomes easier.

If these absolutes were true, and there is no value to be found among specific races, then point to the race that exhibits, as a whole, total criminal behavior or universal civility. What race has made no positive contributions to the modern world? The truth is that all races have contributed to Utopia's foundations. As many others have pointed out, economic class tends to affect criminal behavior more so than any other factor (again with no absolutes). It is obvious that impoverished groups are more prone to destructive behavior than other economic classes. Should the rich and middle classes kill the poor because they produce the most crime? What, then, when crime still exists? Continue to target classes to the point of extinction?

There will always be a way to oversimplify and overgeneralize the dimensions of social problems. Using single pieces of statistical data to form conclusions is illogical. For example, the bank robbers wore pants. Therefore all persons wearing pants are bank robbers. These types of conclusions are the furthest points from finding the truth, and thereby finding solutions to the problems we face. We need a vast array of data to form a clear picture of our problems. There is no other way from considering all aspects of a situation to accurately gain insight. Blaming a complex abstraction, such as an entire race, for something is an obviously emotional response, void of conscious deliberation.

Capability (and therefore a potential for positive contribution to society) is gained by those fortunate enough to have its capacity and who are willing to work at developing it. Success can have innumerable contributing factors, although effort tends to positively affect its potentiality, and if you are determined to avoid it, you can certainly do so. All races exhibit these potentialities. Unity should be the focus of our attention as opposed to further defining our capacity for discord. If we throw away any race, we throw away with them all the contributions they will ever make in this world. In doing so you may throw away the cure for cancer, countless lifesavers, great teachers, born leaders, or possibly even the greatest mind the world has ever seen, and all of this so you can live in a world of less diversity. Can we afford such waste as a society?

The racists of the world have no need for foreign cultures, because they are content to never see the beauty in a world of complexity. They would rather live in a smaller world, one the size of a single city as opposed to the thousands of miles of assorted terrains we share. They would trade any amount of progress to remain comfortably protected from an irrational fear, a fear that perhaps their ancestry alone does not entitle them to some intangible birthright. In their eyes, their failures can be blamed on others. Their shortcomings and lot in life are someone else's fault. In some cases it is the fault of those they have never bothered to attempt to understand, people with just as much potential and value as anyone else, people who they've never met and who've had almost no impact on their lives. These so-called monsters could just as easily be a myth, and still the racial extremists would waste their lives in search of them so as to have someone to blame and destroy for their miseries, many of which they would not have if they looked for the positives in life instead of dwelling solely on the negatives. What kind of a wasted life is this?

All races have heroes and villains. Racism is just another of the many things slowing us down. Why can't we get past such pointless obstacles and work together to build a better world? Why has this, of all things, after centuries of progress, not become the object of universal ridicule it deserves to be? Since its root is excessive pride, perhaps the answer to its long overdue departure lies there. Racial pride and its cousins, national pride and gender pride, need to be reevaluated, the core question being, is it justifiable to take pride in your ancestry, your nation, your gender, or any other variables of purest chance? Let's focus on them in turn. The first is the easiest, most of us can agree on that one, but I believe all are equally unjustifiable.

By what right do I take pride in something that someone whom I've never even met has done? I had no hand in their struggling through hardships, their survival, their defeats, or in any of their successes. I had no input in any of their decisions and no part in any of their crimes and mistakes. It is an obvious case of unjustifiable pride to say that I'm proud of these people who could have been kings or thieves for all I know. What difference does it make in my life? These people, to whom I'm referring, are my ancestors. If my distant ancestors were royalty, then what good does that do me here and now? It doesn't change who I am. How can I possibly take any credit or guilt for their actions or reasonably assume that since they were great, then that must mean I'm great as well? By this same token, I have no right (reasonable justification) to use their lives for an excuse, such as, since they were alcoholics, it's all right for me to follow their lead. How about, since my ancestors were failures in life, that means there is no use for me to attempt success? Does that sound right? It is a flaw in logic so base it could be called delusion. At the very most, I can be happy (not proud) about the fact that my ancestors did have sex since that led to my being born, but should they have received a medal for this astounding feat? They managed to keep their progeny alive, to some degree at least, though even this could have other explanations as well. Some were probably orphans.

We hear statements such as "My ancestors were warriors," and we should cringe. Congratulations are not due for having been born, though the event should be cause for celebration. Having powerful, brave ancestors does not make you powerful, and it doesn't mean that you're not a coward. It just means that a long time ago some people had sex, and here some are boasting over it. Until, as a society, we learn to be strong for ourselves instead of just for dead people, we will remain in part as parasitic scavengers. How much more impressive would it be to be able to say, "All of my ancestors were self-destructive losers who wanted their children to fail in life the same way they had failed, but I strived hard at something and managed to use it to work my way out of a regrettable situation"? Who's the warrior now? Who has worked for something, overcome obstacles, and earned their attainments?

Saying your ancestors were great is like saying your ancestors were exceptional, which is like saying that they were better than other people, which is like saying that your people are better than everyone else's people, and that makes it racism. No child should ever be led to believe that because they have been born to a certain race they are incapable of attaining what someone else has accomplished. These are simply false assertions. It is wrong to discourage others, and by taking unjustifiable pride in something, this is exactly what you are doing.

It may sound like from this attack that you can't take pride in anything, but this is not the case. You can take quite justifiable pride or shame in the things that you have had a part in doing. If through your efforts (actions or non-actions) you cause, or help to cause, an outcome, then you can take the appropriate measure of pride or shame in the results. If, however, your existence has had no effect on the outcome, then you have no right to take any pride or shame in the results.

To clarify, children cannot take pride in the actions of their parents, no matter how great or loving and caring, since at best their simple existence (their parents' desire to provide for them) was the only part they could have possibly played in the said outcome, for good or ill. Existence is not something you can take pride or shame in. However, parents who provide for their child can take a measure of justifiable pride in the accomplishments of that child if, in fact, they in any way helped them to accomplish a task (such as by providing the means for the accomplishment or by teaching them). They cannot take justifiable pride in simply having reproduced. If you helped to do something, you can take pride in your part alone, not in someone else's. Also, they could find justifiable shame if they were to harm their child's progress.

What about your nation? Can you take pride in that? You can only take pride in the parts to which you have contributed. Let's use a typical example followed by an extreme example. Is obeying the law and paying taxes enough reason to take pride in my country? No. No decent measure of justifiable pride can be taken from these bare minimum actions. Giving up tax dollars buys the services your government provides. Taking pride in a birthright (natural citizenship) is an attempt at laying a claim on something for which you haven't paid. One could pay their taxes and still hate their country. They could possibly hate it for taking the taxes in the first place or for using the funds poorly. One could obey the law reluctantly out of self-preservation as opposed to an admiration for civility and order. Being a responsible, civilized taxpayer makes you of use to society, but if this is all you've done, you have not contributed to the government that serves society.

Now let's look at an extreme example. Imagine for a moment you are a soldier who is present at one of any nation's biggest disappointments, say, for the sake of an example, a massacre of innocent people by a country's military. Imagine that you stand in this moment of monumental decision. You may either go along with it, killing the innocent, flatly refuse, attempt escape, or attempt to prevent the murders from happening. These are the general options available to you, and any of them could carry heavy consequences. If you killed the innocent, then afterward you might feel justifiable shame as well as experience the justifiable contempt of those you are sworn to honorably protect. You could then attempt to justify your actions by telling yourself that anyone would have done the same thing had they been in the same situation. However, if you had no part in the event, then you could feel no justifiable personal shame or any justifiable contempt from others. The sole exception to this could be, if you were to admit to yourself that you would have done the same thing the others did, lacking the courage or common sense to abstain from mass murder. Then you could feel some vague sense of valid shame, but still, no reasonable blame for the crime. If you were not faced with the decision, you could not take justifiable pride, shame, or blame in its outcome regardless of how great or disgraceful, since you were not a contributor. It's easy to say or to imagine that you would have done something in a given situation, but talking about it, imagining it, and doing it are completely different things. If talking about it or imagining it were equal to doing it, many of us would be the most courageous and heroic of figures while at the same time being the lowest and most contemptible.

Each and every time you talk about how wonderful your country is, you cannot help saying that every other country in the world isn't as wonderful as your own. If you say, "We're number one," what does that make the rest of the world? By saying a system of government is the best, you can voice your opinion about the system, but to say that your overall track record is the best is childish. The worse is to say, "I'm proud to be a so-and-so." It is only by chance that you are a so-and-so unless you've earned it. You could have just as easily been born anywhere else and would possibly be chanting the same slogan, but in regard to a different nation and in another language. Because you happened to have been born on a specific tract of land that was inside a boundary is no accomplishment. If, however, you were to earn your citizenship by means other than birth, you can take pride in the work done to achieve your objective. Some of the United States' greatest patriots have been immigrants, but even they have had no right to take pride in their nation until they contributed to it to some degree. The rights you have are to be cherished, demanded, honored, and if necessary, fought for to secure. They are not a reward for managing to be born "over there" instead of "way over there."

"Well, can't I take pride in the fact that ancestors provided my rights?" No, you cannot. You didn't fight for them or work to define and establish them. You aren't the one who had to risk your life obtaining them for yourself and for your descendants. What you can be is perhaps better than proud. You can be grateful and respectful of your birthright. You can *value* it. You can learn from the past and try to honor the memory of those who made the right decisions, not try to steal a piece of their accomplishments for yourself using the chance event of heritage as your weapon. If you've made a contribution, then you can take your measure of pride and decide for yourself whether or not to be patriotic by putting that pride on display. Until then, to wave the flag while doing the bare minimum is, in my opinion, a blasphemy of what it represents. A thing cannot be truly revered while at the same time taken lightly. It is not the banner of a sports team, even though it is too often used as such. It would be better to demonstrate your freedom by burning it than to put it on display without sincerity or consideration. The great people who have made contributions to making your country a better place would probably like for you to do the same, instead of just chronically symbolizing that you do the same or would if a need arose. It begins to look like insecurity. There is nothing wrong with simply treasuring your advantages. You don't have to turn them into something they are not in order to properly appreciate them.

Now for gender, you can already see where this is going, can't you? Just as by coincidence you were born in a particular place, so you were born with a body. Possessing one does not make you special, because we all have one. Both genders are needed, and both add value to the world. Possessing a body that is more naturally capable in some regard is no achievement. You can value the circumstance of birth without using it to change luck into false accomplishment. If you work on your appearance, health, strength, knowledge, etc., then you have a right to take a reasonable, appropriate measure of pride in the results if you choose to do so, but natural gender identification is not, in itself, an accomplishment. Regarding sexuality, only if there has been a conflict, which you have been able to resolve, have you accomplished something. However, this ends up being its own reward, and little in the way of capability has been developed. People of both genders have furthered Utopia's cause. We should work together to reach our potential regardless of our differences. Sexism, like racism, is an archaic waste of time. We can all add value, and we need each other to achieve our full potential. We should be working together rather than childishly pointing out our differences in such stereotypical detail as to only further incite discord.

A few words now on gay pride. While being completely sympathetic to the cause of equality of rights, and social tolerance among all people regardless of race, nationality, gender, or sexual orientation, I can't say that I would approve of someone claiming they are proud to be homosexual any more than I would approve of someone claiming they are proud of being heterosexual. It sounds far too much like "I'm proud that I like ice cream" or "I'm proud of not liking coconut." It would be more effective to say, "I'm not ashamed of my sexual orientation." Some factions of the civil rights movement had a similar rallying cry, "Proud to be black." In a world of individual responsibility this must be seen as being just as factually erroneous as "Proud to be white." Otherwise you are saying that when a crime is committed, the entire race and/or nation of the perpetrator must be held accountable. These sentiments obviously still exist to some extent today, in this, as well as in other countries. Should we really be proud or ashamed of these trivialities? Perhaps it would be better if we reserved words like "pride" for true accomplishments rather than such relatively insignificant aspects of our lives. Finding your desired sexual orientation is no greater feat than finding a taste or dislike for anything else. It is a temptation in a repressive society to jump to the conclusion that, because you treat me as if I should be ashamed, I will declare the exact opposite. This is an overreaction. It is responding to a prejudice by becoming prejudiced yourself. I can honestly say that if placed in such unfair situations as others have experienced, there have been many times in my past when I would have probably had a similar overreaction to such derogatory treatment. However, this rhetoric evolves into pointless bickering. My race did this and that. People of my sexual orientation have done this and that, as if keeping score was as important as putting aside these nonsensical squabbles that continue to separate us. They serve to obstruct Utopia's formation.

One person says, "I'm proud of my ancestry and taste."

Another responds, "Oh yeah? Well I'm proud of mine, and it's different from yours."

"Mine's better."

"No, it's not, mine is."

"But, mine has done this and that."

"Yeah, but yours also did this, and we did that."

And on and on with more of the same…

This has been going back and forth for the last several thousand years while, on both sides, needless suffering continues. Such arguments are emotional, rather than logical.

Potentiality (for good or ill) is the same for all races, genders, and sexual orientations.

There are those who have been oppressed over their race as well as for their sexual orientation. There have been those who stood up against the bulk of an unjust, ugly society to say, "We demand equal treatment." These people are not the heroes of the black cause or of the LGBT cause. They are heroes of a much more important cause. They are the heroes of freedom and equality, something all can appreciate. They are Utopia's heroes. All those who stand up in the face of tyranny, regardless of how overwhelming or trifling it has been, have made a contribution. These people may take quite a lot of justifiable pride in their hard work, risks, struggles, and triumphs. We enjoy much (far too often taken for granted) that we owe to the combined efforts of those seeking justice. Whether we acknowledge it or not, we do want a society free from senseless, worthless prejudices. We want the world of harmony we are slowly gravitating toward, as opposed to a bigot's fantasyland. Utopia has no time for such prejudices. It is too busy trying to find the next best way to improve all our lives to give concern to inconsequential matters. It doesn't care about the ethnicity, gender, or sexual orientation of the persons who finally discover the cure for cancer. It doesn't matter. All that matters is that the cure is found. The only time spent on the subject in Utopia would be a moment's reflection on how nice it is to live in a world where we have the opportunity for individuality and choice, a world where we can take advantage of our widespread diversity instead of cultivating narrow-mindedness and uniformity.

If you have stood up against these prejudices, then you have caused an effect worthy of some sense of personal achievement (justifiable pride) since you have changed the world for the better. If you have come out as being gay, for example, and risked losing your loved ones in the process, then you have accomplished something, though part of your reward is that you don't have to be secretive about your lifestyle. You've made a statement. Statements can have great effect, but only when there is a reason to make them and only if they are worth the effort. Individuality is no great accomplishment. It's what you do with it that counts. If you have gone a step further than simply making a statement, by petitioning the government or society for equal treatment regardless of ancestry or a value of taste, you have encouraged positive change. You can still take no justifiable pride in being gay. You can take pride in something better. You can take pride in having the courage to face difficulty and stand up for what you believe as being right. This is something we can all benefit from and enjoy, including the future generations.

The crux of this entire argument could be rephrased as "When is it justifiable for me to feel a sense of pride?" the answer being "Only when you have affected an outcome." Ask yourself and answer truly, "What changes have I caused? How did my actions or nonactions affect the outcome I wish to embrace or renounce?" You can find which things in your life to take pride or shame in using these questions as your guide. Everything else is external to the question.

As I mentioned earlier, by what right do I take pride in something that someone whom I've never even met has done? The only thing that I could possibly take isn't pride or shame; however, it is nonetheless useful. What I can take is encouragement or disappointment from others. We can learn from their mistakes and successes. We can say, "My fellow human accomplished this great thing, and since they accomplished it, then so can we all learn from it, appreciate it, and know it is possible for others to accomplish similar great things." We all have an equal claim to this encouragement and disappointment regardless of race, nationality, gender, or sexual orientation.

Throughout this rant, I want to make one point absolutely clear. Do not throw around the word "pride" without due consideration. It is easy to take it lightly or measure and allocate it improperly. By that same token, do not underestimate the value of words like "admiration," "happiness," "sadness," "respect," "contempt," and "disapproval." Do not use them interchangeably with "pride" and "shame" when another would serve better. Pride is useful for validating and affirming beneficial ends. It is dangerous when not absolutely justified (i.e. when no good can come from it). Shame is useful for avoiding past mistakes. It is dangerous when dwelt on to the extent that it produces self-hatred. Blame is only useful as a mechanism of justice. It is dangerous when used to incite lynch mobs or as a diversion of personal failure. When blame is thrown around without consideration it usually points to a guilty conscience of the one assigning it. When shame is taken compulsively, it points to an inescapable obsession with the past instead of a usefully learning from it before moving on. When pride is taken unjustifiably it points to a desire for unearned attention. In Utopia people must understand how not only to use pride, shame, and blame to their fullest advantage, but also how to keep them in check.

The Value of Responsibility

"There is a limit to the legitimate interference of collective opinion with individual independence; and to find that limit, and maintain it against encroachment, is as indispensable to a good condition of human affairs, as protection against political despotism."—John Stuart Mill

The freedom to follow or abandon a particular ideology is critical to progress. Choosing our own path and our own moralities are responsibilities with which we can all afford to be trusted. Any who suggest otherwise are looking to chain themselves to an inflexible tyranny incapable of meeting the challenges of a world in motion. Utopia's citizens accept a higher degree of self-rule than is currently the world standard. This means fewer laws, more freedom, and more personal responsibility. How can we improve our professional lives or our personal lives if we cannot be trusted to decide moral issues for ourselves? If you choose to do something dangerous, the ideal state would warn, not restrain you. To effectively prevent you from harming yourself, the state would have to monitor you to some extent. It is not the government's responsibility to babysit us from birth to death and make sure we do everything we can to live as long as possible. The successful government enforces laws, not moralities. This distinction should be made clear. Many see law as based on morality, whereas we would be better off viewing law as that which protects us, our rights, and our property, including any communal property we may share. Morality and immorality should be that which is debatable concerning virtuous and improper behaviors.

Some may say murder is immoral, hence it should be illegal, but by that same token, another person could say gossip is immoral. Are we ready to make gossip illegal? I would say murder is first an infringement on another's right to life before it is obviously immoral and uncivilized. This first part is what sanctions judicial intervention. Morality should have nothing to do with legal issues. Morals vary and are subject to personal interpretation. Laws are ideally more rigid, and should only be subject to judicial and legislative interpretation.

There is great value in making your own decisions. It is an optional weight on your life. The person who values freedom must value accountability. Without personal accountability, there can be no virtue. By legislating virtue, you cheapen all morality since anyone being virtuous is simply fulfilling natural self-preservation. You can take no justifiable pride in simply obeying the law. You can, however, take pride in contributing to society of your own volition. If helping an elderly person who is having difficulty crossing a street were a legal obligation rather than a moral choice, what else could be made into legal obligations? Eventually everything deemed good could become a legal obligation, and people would live their lives absorbed in total attendance to a myriad of laws. In reality, people could not cope with such intense control. Society could not function if you were to add to this list continually, since there would be too much open to personal interpretation. Contradictions would also arise. People could argue to no end whether the elderly person had really needed help or not. We have to be trusted to decide many things for ourselves. I see helping an elderly person having difficulty crossing a street as a virtuous act, but I do not recognize the right of anyone to force this responsibility onto others. The same can be said of any moral issue. I see ideal legalities as defining them in fact. That which is not mandated by law, but benefits society is virtue. Enforcing moralities destroys charity. It would cease to be charitable if it were mandated. The government removing itself from such matters as it cannot positively effect would be advantageous to both the system and society.

For the vast majority of human history, morality has been the province of the world's religious organizations. Usually the only justifications for their contradictory moral interpretations are the claims of their ancient texts. The real answer as to where morals came from is that they came from humankind miraculously developing the intelligence to discover their value. Nobility is a byproduct, and immoralities will eventually be recognized simply as useless behaviors. In a society where the freedom to choose religious beliefs is enjoyed, there too should be the freedom to choose our own principals and moral codes. In political terms, a religion could be said to consist of a moral code reinforced by spirituality or mysticism, depending on where you are standing.

A defining characteristic of justice is fairness. The ends of a system of justice are our protection and retribution from harm. Its procedures and machinations are designed to best facilitate a civilized, structured society. Providing justice perfectly is an impossible task. I feel overwhelming respect for an institution attempting such tasks and providing us with such benefits as the justice system endeavors. However, I do want it to tell me I can choose my own morality and be responsible for my own life. I want it to give me the freedom to make bad decisions and to learn from them. I want it to give me the freedom to fail just as deeply as I want the freedom to succeed. I want to relieve it from the burden of any responsibilities it may feel it owes me over the personal consequences of my actions. I do not wish to ever trouble a court of law with my own blunders that cause no harm to others. The justice system provides such an incredible benefit to our lives that seeking to hold it, or any other entity, responsible for our well-being, to unrealistic and unfair degrees, forgoes our own responsibility over our lives and is wretched beyond redemption. I want the government to place society's imperatives on the shoulders of society, where they belong, and stick to their own province. I no more want the judicial system to dictate our morality than I want a lynch mob to take on judicial matters.

We are missing rights we could be using to aid in the development of a better government. We should, to a greater extent, be given charge of our lives and should face these new responsibilities maturely. We are entitled the right to shoulder our own burdens, to accept the consequences of our actions, and perhaps most important of all, we are owed the freedom to fail. Without the freedom to fail, there is no freedom, or any of the blessings of liberty, mentioned in the Constitution. Without the freedom to fail, we are being forced to succeed without any definite knowledge of, or appreciation for, how to succeed. Should we be told what to eat or how long to exercise in order to save our lives and benefit our health? Should we be told what to wear for our mental well-being? By doing so, you would be creating slaves who are being taught from a young age that they are the freest people on Earth.

Where should these lines between governmental responsibilities and societal responsibilities be drawn? The more we restrict the government to its proper sphere, the more effectively it will provide its functions. The rest is up to society. Hand in hand we can build a better future. Can you imagine a better system? Can it hold up to the scrutiny of the masses and the tests of time? Can it prove itself to be better than other models? It cannot be built by an individual. It can only be built by generations of those seeking to improve it. The following are some adjustments that may be of benefit to society. Like Utopia, if they really are the better paths, then they are inevitable, and the sooner we adopt them, the better. They are highly debatable, sensitive subjects for many, and as strongly as I do think the nation/world would be a better place with these alterations, Utopia may still prove me wrong.

The Functions of Rule

What is the proper role of a government? What is its purpose, its job? What things should it concern itself with? All can agree that it exists for our well-being, but to what extents should it go, and what truly is for our well-being? Too much intrusion is as harmful as not enough. It attempts to provide for the greater good in all things, and the greater good is quite difficult to define. The questions over where these governmental lines are to be drawn have caused endless debate over the ages.

Two important factors of the greater good are justice and freedom. They have an inverse relationship, yet depend on each other to some degree to provide benefit. Absolute freedom would mean zero justice. Absolute justice would mean zero freedom. Neither extreme would work to society's benefit. Somewhere in between hides the ideal state. Locating it is one of the many challenges humanity faces. The task is daunting. Once found, it would not be instantly apparent that it has been gained under any current governments of the world. Only time would prove its success. As we near the ideal state, our methods for monitoring progress will improve, and we will gain a deeper insight into what works best.

To make matters more complicated, any evidence must be taken into such broad consideration that finding what works and what doesn't can be exceptionally difficult. Just because a system of government or a facet of it has succeeded or failed in the past does not necessarily mean it is the correct or incorrect method since there are so many contributing factors to governmental success and failure. The right path could be taken and still fail due to other circumstances. In light of this, how can we determine anything conclusive about our ideal directions? The answer is the same way that we've managed over time to map the sky, through meticulous observation and documentation. We must study this intricate, ever-changing picture of our political and social environment and try to take in the maddening number of effecting factors. Betterment can thrive once a more thorough understanding of the landscape is achieved.

Let's take a look at the highest legal document in the land, the Constitution. The opening paragraph is short for a document of its size and outlines the basic functions of the new government:

"We the People of the United States, in Order to form a more perfect Union, establish Justice, insure domestic Tranquility, provide for the common defence, promote the general Welfare, and secure the Blessings of Liberty to ourselves and our Posterity, do ordain and establish this Constitution for the United States of America."

The functions broken down could be categorized as the following:

- Establish justice
- Provide safety (domestic peace and national defense)
- Promote the general welfare
- Secure freedom

There are probably few credible arguments against the government providing justice and protection from threats (internal or external). All countries provide these services to their citizens in varying degrees, though by what means they provide them are sometimes called into question, along with their accuracy and efficacy. Freedom is provided to varying degrees, with some countries going so far as to avoid mentioning the subject, since they guarantee and provide so little of it to their citizens. Freedom must be traded off if these other three areas are to be taken into consideration. We are not free to murder in the name of justice and the protection of rights. We are not free to dump trash in the streets in the name of the general welfare. We are not free to buy a rocket launcher in the name of safety. Since there is a trade-off in each case, with each fighting for the greatest value, we must ask, in regard to every issue we face, how it will affect the causes of justice, safety, general welfare, and freedom. For example, the right to consume alcohol (or any intoxicant) has a detrimental effect to safety, a positive and negative effect on general welfare (economic gain versus collateral damage from misuse), and a favorable effect to freedom.

Has the *promotion* of the general welfare become the *providing* of the general welfare over time, or does it need to be promoted more than it already is? What will yield the greatest benefit? Should general welfare ever be considered paramount to justice? To what extent should it take precedence over freedom? Do its considerations know no bounds? General welfare could get out of control easily since any number of breaches may be committed in the name of the greater good. People could even be killed wholesale in the name of the greater good, with little concern for justice or freedom. What is in our best interest? Some say we should allow the government to plan for our basic financial needs once we reach a certain age. Some say since we are capable of doing it ourselves, we should be freed from mandatory participation in social security programs. Again, where do we draw the line? To what extent should our government protect us? Killing off or conquering all foreign nations would eliminate all external threats, but is that sensible, rational, helpful to the cause of freedom, a promotion of the greater good, or just? To what extent should we be free? Should we be free to kill or otherwise harm ourselves? If not, then what happens when two doctors disagree over the appropriate treatment for a physical or mental condition in a patient? If health were mandated by law, how would the government decide which doctor is correct? To what extent should our welfare be promoted? Should all homeless be eliminated to promote general welfare? If so, then by what method, killing them off, supporting them with a new universal welfare system, or building a better world where poverty is minimized? Do you want your tax dollars to fuel death and unnecessary imprisonment, to support the masses regardless of their efforts to provide for themselves, to keep the worst criminals alive in the name of justice? To what degree should we be free to lie or swindle? To what degree must the buyer beware? Should we be free to make obviously bad decisions and fail in our endeavors? And what about the method for taxation that provides for all of this structure? Flat taxes, sales taxes, income taxes, state taxes, property taxes, which method or combination of methods for collection are the fairest? Which are the most efficient, fair, and cheapest to operate? What penalty for nonpayment is just?

To what extent does the government involve itself in the economic matters of the nation? Where we draw these lines defines the type of government we have (allow, tolerate, expect, employ, desire, need, or whatever other word you would like to use). Society's role is to expect the proper functions and to see their paramount concerns addressed through representation and through petition. The government should do its job as well as possible and expect reasonable support from its citizens.

Justice

"True peace is not the absence of tension: it is the presence of justice"—Martin Luther King Jr.

A system of justice is one of the most underappreciated things in our lives. Out of necessity it attempts the impossible. It seeks to create an environment where reward and punishment reflect the concept of fairness. Without its benefits, we would have little use for the other three functions of the government. Having said that, I believe there is a great deal of room for improvement in our own system as well as others. Too little is being done to boost its effectiveness and the efficiency of its mechanisms. Its own parameters of operation often interfere with its ultimate objective. Courts get backed up to absurd lengths of time. Prisons are overcrowded to the point that dangerous criminals are released into society. Major crimes go unaddressed. Perhaps worst of all, there are laws in place that exist only as a way for the system to protect itself from potential criticisms. Certain laws, such as overly low speed limits, for example, attempt an unreasonable amount of safety so as to protect the lawmakers from claims of neglected security. It turns everyone on the road into a criminal and presents police with an unfair dilemma. At some point the system should vocally expect people to protect themselves. It should address all valid concerns, yet have no fear of unjustifiable public outcry. Otherwise, we could sue the government for not enforcing exercise routines, or restaurants for using too much fat in their products. We could sue the hammer maker for an accidentally smashed finger. We could sue the power company for electrocuting us when we stick our finger in the light socket. There would be no end to it. None of this is justified if we are to have any responsibility over our own lives. We should not want everyone to be liable for our safety to an unlimited degree. If they are, then they have every right to control our lives to an unlimited degree. We should want more responsibility, not less. These potential criticisms are feared, regardless of their validity, by the one institution in our midst that should, above all others, define what justifiable offenses are in the first place.

The law is one of our greatest assets in life. We have all enjoyed its protection from birth. It separates us from a life of chaos. Without it there is no hope for progress. The law should be respected and loved, yet it is far too often made into a laughable subject and in many cases justifiably so. Laws are written that are hopelessly unenforceable, making law and justice into a disgrace. Other laws are seen as optional, which violates their definition. Complex cases are thrown out instead of addressed properly. These situations must be remedied. If I can make only one additional assertion to the three basic principles proposed by this writing, it would be: one of the greatest hindrances of betterment is the outright lies being perpetrated on the youth of our country. Until we stop lying to the future generations, we will continue to sow feelings of hopelessness in society. The young are taught to see the government in one aspect throughout their school years, and learn that the realities are otherwise when they face them firsthand. Until we start making good on our most absolute declarations, there will remain a hindrance to the enthusiasm of those who would otherwise wish to join in the age-old cause of betterment.

In an effort to prevent the possibility of an innocent person going to prison, great measures are taken in the processes of justice. Almost everyone has heard "It is better for ten guilty persons to escape than for one innocent to suffer." In light of the awesome responsibilities of the justice system, this is an admirable and effective stance. It is more intrinsic to the system than many know. Several procedural rules exist due to this sentiment. It is a sentiment as old as the first fair trial of a person by a society's representatives. These rules often let criminals go free, but the immeasurable benefit is that fewer innocents suffer. Nonetheless, the judicial system should audit itself. It should take note of the instances where justice was not done due to the stringent rules that seek to protect the innocent. When an obvious breech occurs, the issue should be addressed. Periodic and situational judicial reviews should be exercised to further the cause of justice. All valid concerns should be addressed so as to minimize criticism and encourage public support for this much needed service.

There is another wonderful saying of great pertinence on the subject that cannot be observed intensely enough: "Fiat justitia ruat caelum." (May justice be done though the heavens fall).

Security
"Well may the boldest fear and the wisest tremble when incurring responsibilities on which may depend our country's peace and prosperity, and in some degree the hopes and happiness of the whole human family."—James K Polk

"What then of war?" One could ask. The purpose of the war in question would have to be determined in order to answer properly. There is also a notable exception to such a strict definition of murder called, the concern for others' safety. In applying this exception to nations, as opposed to individuals, much must be taken into consideration, such as the primary responsibility a government has to its own citizens before others can be addressed, the sovereignty of the involved parties, the situation itself, the legality of whatever action is being considered, and the possible outcomes.

This is the grave situation our national defense attempts to confront on both the local, national, and international level. To what extent we govern our own affairs varies greatly from the question of how far we interfere in the affairs of other nations. The US interfered in World War II possibly just in time to help prevent Hitler from developing atomic weaponry. Granted, this was after being attacked by Japan, which, along with Germany's declaration of war, finally initiated full scale involvement (there had been indirect support for the Chinese and British before these events). We have no choice but to keep an eye on world affairs to ensure our own safety, though US international intervention has, at times, been widely considered far too zealous. Where do we draw the line between preventing tyrants from gaining a dangerous level of strength, and overly meddling in affairs best left to the nations in question? International relations are a major element of our safety. It needs a high degree of attention, definition, clarification, and refinement if we are to maximize our efficacies.

One child builds sand castles, while another goes about kicking sand castles others have built. Some value construction, while others lack this value and make up for it with deconstruction. Others still both build and destroy. Society is worth maintaining and defending. Civility demands we seek peace, yet it also sees great value in what humanity has managed to construct, both physically and socially. We have many more options today than in ancient times. We can have more still if we choose to work smart in social endeavors as well as in our professional lives. To build this world, security from harm is needed, and to facilitate this gain, a certain degree of authority is required.

It is important to note that if a country were attacked on a massive scale with the intention of its total annihilation, and it began to break down, safety would gradually become the paramount focus. In the worst of times, everything else would become secondary to this primary consideration. Welfare is of little use to the dead (though it could still be of use to their loved ones and progeny). As total destruction neared, many current considerations would be abandoned entirely out of expediency. It would be of use to have more clearly defined values to which we hold fast to the very end, regardless of a hypothetically approaching extinction. This code would not only be of use to a specific nation, but also to the world as a whole. This would include the facets of the previously mentioned, universally agreed upon form of ideal justice. To form solid principles we must establish, as clearly as possible, that which we will observe to the absolute end.

For example, torture is widely considered to be justifiably illegal in almost all circumstances, as it should be, but we should also consider the worst possible scenarios to form the determinations on whether or not it should be used. This is typically referred to as "the ticking time bomb scenario". It is the hypothetical situation that if a terrorist group or individual planted a massive weapon in a major city or cities, and the organizers of the plot were captured before the weapon's use, would torture then be justified to attempt gaining information as to what city or cities should be evacuated? When do concerns for humane treatment of suspects or prisoners break down, if ever? If it could possibly save millions of lives, would it be justified?

These types of extremes are what should ideally help determine our approach to our principles in this regard since they help to define the issue. Meaning, the court should be as prepared as possible for whatever circumstances it may face. Principles should be defined as, "that which we will absolutely never abandon" not "that which we observe when convenient". Laws should not exist only to be excused by the extreme situations that the tests of time may yield. If they are to be excused, they must be specific and clear about what they excuse, such as when homicide is deemed justifiable. We should hold our military leaders accountable for their use of force, but should also probably take the situations into consideration as well, rather than trying to establish too many absolutes regarding what may be the most complicated of all endeavors, trying to provide safety in as civilized a manner as possible. This effort is further complicated by the fact that the enemies of true civility are often extremely uncivilized and operate by an entirely different set of standards.

What I am calling for here is not a universal, set-in-stone, yes or no call on torture. I'm suggesting it may be more advantageous to consider these, hopefully rare, circumstances on a case-by-case basis rather than looking to condone or condemn them wholesale. There is no question that regularly sanctioning such practices would corrode the system of justice. Delegating the authority to the president is no solution either, for at least two very good reasons. Firstly, giving the president such authority takes us closer to a dictatorship and robs the judiciary of its responsibilities. The president is not a court of law. They should not be expected to do everyone's thinking for them or administer justice. It is disgraceful to expect such misdirection of authority. The second reason is that in the ticking time bomb scenario (one of the very few justifiable arguments for the use of torture), by definition, there is not enough time for such protocol as decision making by a high authority.

Remember that the idea is to protect our protectors while holding them accountable for dishonoring their positions when applicable. Imprisonment could in itself be considered a form of legalized torture, and yet the justice system is trusted to handle its parameters. Consider the abuse many prisoners endure in the country's worst prisons. All punishments could ultimately be argued to consist of various forms of torture. Putting hermits in the general population would be torture compared to placing them in solitary confinement. The rights of the accused should be held to as high a standard as possible, and torture probably brings about poor results and unforeseen consequences in practice when compared to other forms of interrogation, such as those that exhibit positive reinforcement.

Laws should be clearly established regarding the general whens, wheres, and whys. But in the case of national security, it may also be helpful to alter the current protocol so that in each and every one of these circumstances (which should be exceptionally rare if ever), the alleged perpetrators of the torture are made to stand before a federal court to detail and justify their actions. Illegalize the practice to minimize its use, yet allow those responsible for providing safety to explain their actions. This may be a very effective form of control that takes everyone's concerns on the issue into due consideration. Since attempting to outline the infinite number of circumstances is literally impossible, let the law define the scope, and as it does in other such complex situations, let the justice system decide what is just. It may be imperfect, but it is still the optimal tool for determining justice. In a word, I'm calling for *accountability*. The people in question are sworn to honorably protect us and our rights. So long as they do their duty with honor, they should have nothing to fear. Since torture should be discouraged, this would discourage it as much as it can be. The level of tolerance would then be set as with other potential crimes. If we are to conclude that there is never a circumstance that would call for torture then we must understand that this decision may carry consequences, just as all rights do. These consequences may consist of the blood of the innocent, something not to be taken lightly. If, on the other hand, there are to be certain, extraordinary exceptions, then we should allow the courts to perform the function for which they are designed and understand that the unavoidable imperfection of these tools of justice may lead to tolls every bit as serious as the innocent blood that is sought to be protected. At any rate, methods of establishing safety such as torture are destined to fade in the face of an ever-developing world, eventually becoming marginalized by a lack of need for such archaic measures.

Regardless of this issue, the killing of noncombatants is not a civilized means of warfare for any situation. This is a concern of our military, but it cannot be pursued vigorously enough. Otherwise, we negate their positive influences on world peace, such as disaster relief efforts. Though it may be unavoidable at times, siege warfare is an especial inefficiency in any age and should thus be avoided whenever possible.

"Ultimate excellence lies not in winning every battle but in defeating the enemy without ever fighting." This quote from *The Art of War* was probably meant in the tactical sense (such as encouraging ambush tactics), but nonetheless, the most successful revolutions are of the least violent variety.

General Welfare

"Every art, and every science reduced to a teachable form, and in like manner every action and moral choice, aims, it is thought, at some good: for which reasons a common and by no means a bad description of the Chief Good is, 'that which all things aim at.'"—Aristotle

Promotion of the general welfare is a relative term. It would be easy to demonize the concept without pointing out its benefits. Much harm has been done in the name of general welfare in the past. History is full of examples of using general welfare as an excuse for obvious wrongdoings. However, the reason we are not allowed to disturb the peace or adversely affect communal property, among other obviously reasonable restrictions, is due to the concerns of general welfare. There is benefit in such restrictions of freedom. As a society, we must share certain spaces and certain properties. We have a right to certain communal property and resources since our tax dollars have collectively provided it. Such societal welfare issues must be addressed, but must also remain in check. They must never exceed their boundaries. The considerations of all four governmental functions must be carefully weighed with each step we take, and we must constantly reevaluate our past decisions' effects on them.

Freedom

"Liberty is not a means to a higher political end. It is itself the highest political end."—John Dalberg-Acton

Freedom facilitates choice and encourages responsibility. To use an example, take smoking bans in restaurants. Firstly, public smoking bans should be allowed since communal property should be voted on by the community as to how they want it kept (free from clutter, visually appealing, healthy, and so on). However, a restaurant is not communal property. It is private property. No one is forcing you to buy their services. We have taken a wrong turn by imposing our desires on others. The correct course of action is not to petition the government to take away our rights or responsibilities. The correct course of action is to vote with your dollar. If enough people voted with their dollars, the restaurant owners would change their policies or else go out of business. Write to the owners, e-mail them, and communicate your concerns as a customer. Do it as a group, if need be. This would be constructive action, voicing your concerns. They want your business and will listen to their customers. If smoking is legal, then a restaurant providing services to smokers should be allowed to make its own policies. Only in an oppressive society would this freedom be taken by force. We can do more with freedom than with governmental mandates. Personally, I would only want to eat in a restaurant free from tobacco smoke, but I don't want to use the government to take away the rights of a business owner. We need the freedom to fail. We are not owed the right to eat in a healthy, pleasant environment; we are owed the rights of choice and freedom. We have the right to build a society that chooses to create pleasant, healthy environments without using threats to accomplish our goals. If we can force a business owner to provide this environment, then we can force other more unrealistic expectations on one another. We can force behaviors until we are slaves to debatable ideas concerning progress.

Society does not owe us the right to tell people what they can and cannot do with their commercial property any more than with their residential property. This is a small step in the direction of tyrannical control. It is a step away from freedom instead of toward it. Businesses free to choose their own policies will facilitate our wishes without the use of force, and give us options instead of restrictions. This would be the more civilized path, and civility has proven to be the ideal method of change so many times that there is no need to point to a specific example. The evidence is too numerous to even catalog. We are supposed to be proving freedom can work instead of proving we must be controlled to produce benefit. In today's world, more than ever before, businesses are respondent to public opinion and outcry. We must take advantage of this progress instead of forgoing it. If I want to open a business and totally disregard people's desires for certain types of services, products, or environments, then I should be allowed to, and when it fails, I've no one to blame except myself. The business owners should be allowed to control the outcomes of their endeavors since, if their efforts fail, it is they who will suffer, and if they succeed, it is they who will benefit. Obviously, a restaurant should not be allowed to poison its customers, but at some point, forcing them to consider health concerns using governmental mandates breaks down into detrimental control. If there were no end to it, restaurants could all become identical health food vendors, all due to perfectly valid health concerns, but without any regard for self-rule.

I'm not suggesting it will ever reach such an unrealistic pitch, but consider the perspective this one point could bring to other, more critical issues. The USDA and FDA are excellent examples of the blurring of this line between self-rule and governmental rule. Food inspection, for instance, could theoretically be handled by society rather than the government. If a manufacturer fails to provide quality and customers become sick as a result, then the manufacturer would not be expected to remain in business very long. Also, it should be considered that the inspectors could be private enterprises rather than government employees. There is also something to be said for the more socialist argument. The threat of force that regulatory agencies provide (the ability to punish with fines or forbid their practices) may be of use. Perhaps we do need something as strong as government dictation to enforce certain measures in the name of the greater good. Though to save on tax dollars, society would have to step up its efforts. Positive and negative results will be attained either way, and neither system will ever effectively prevent all unsafe handling and processing. Both methods of control present advantages and disadvantages.

A political philosophy attempts, among other things, to draw these controversial lines (where governmental rule ideally stops and self-rule ideally begins), but society should also have its own concern over their application and keep watch for the greatest inefficiencies, since society influences and, to some degree, dictates governmental action. How can we build a better world if we must use force to progress and compulsively look to the government as an answer to all of society's proper responsibilities? Do we really want to look back one day and say we were dragged to our potential kicking, screaming, and under the gun, or that the only way to preserve progress is with governmental control? Being forced to live better will not bring progress. There will always be an imperceptible and largely ignored loss to society's ability to make its own decisions. What good is progress if to gain it we must be enslaved?

Another area where freedom could be performing social betterment rather than governmental dictation is in the
area of discrimination and labor laws. If a person wants to
run a business with stated racist hiring principles, then let
them do so. It is not up to the government to shut them
down or regulate them. It is up to society to shut them down.
Could any business function by establishing stated discriminatory hiring principles in today's nation? No one would
want to do business with an organization practicing such
suicidal procedures. They would be avoided like the plague.
The vast majority would alienate them, and their competi-
tors would thrive.

Racism is dying away, and for the government to set
demands on a company to hire minorities is in itself a racist
act. The very idea of protected classes is a concept of prejudice. The numbers of your class should not entitle you to a
job over another person. I don't blame anyone for taking
advantage of this situation since we are all trying to pros-
per, but I do blame the government for selective favoritism
based on inconsequential values rather than reasonable justifications. Giving any class a government-backed advantage
creates a prejudice by definition. If you own a business, you
should be allowed to hire the best person for the job regardless of race, gender, or any other considerations. You
should also be allowed to hire people based on anything
you choose. This would bring about greater social aware-
ness and responsibility. If you only care about your business
succeeding, you should have the right to pursue those objectives unhindered and unbiased by class mandates if you
choose to do so. Likewise, if you want to employee only certain minority classes, you should be allowed to. Fairness and
equality between the races and sexes should not mean forc-

ing businesses to hire certain percentages of certain classes, giving business owners yet another hoop to jump through. It would be admirable, respectable, and socially advantageous to hear all minority classes claim universally, "We do not need or want special treatment or attention. We have a right to be freed from abject favoritisms and to be treated no differently from anyone else." As to the disabled, no company would want the kind of publicity accusations of discriminatory hiring procedures would bring, especially if it was to save a trifling sum on a minor accommodation. It could destroy their reputation.

Many businesses spend large sums to make sure they are beyond the enforced class mandates. At some point we must stop nitpicking their efforts and let society do its job. These measures at length will become unmistakably condescending. The powers that be seem to say, "Sure you're just as good as everyone else, and to prove it, we're going to force them to hire you." This makes no sense.

Again, the government does not owe us a job. When we look to it to provide us with jobs or hold them accountable for unemployment, we lose the responsibility over our prosperity. If they owe us a job, then they owe us prosperity. In the face of these proposed excessive controls, choice and freedom must shrink. We cannot hold the government solely responsible for prosperity and have business freedoms at the same time. Instead of forgoing freedom, we should be examining how the government is hindering the development of business rather than holding them responsible for providing business. Prosperity must be, to a great extent, society's responsibility. Society has slackened in its responsibilities if we need the government to impose ridiculous, impractical sanctions on businesses to enforce moralities as if they were law. We should band together to make sure our desire for fairness and unity is provided, not tell the government to do our job for us and end up enforcing fairness and unity at gunpoint. By doing this we have petitioned the government to take away our right to build a better society.

This whole problem of legally enforcing racially discriminatory hiring practices in the United States began at a power company, where they were hiring workers based off of educational requirements, and since in that area at the time, black people were largely lacking high school diplomas, a disproportionate number of jobs were going to white people. The problem here was not racist hiring practices. The problem was that a private monopoly existed where people had no right to vote with their dollar. Competition helps to enforce the will of society immensely. Our system attempts to prevent monopolies using antitrust laws, but it would possibly do so more effectively by simply removing the barriers of those wishing to enter monopolistic markets such as utilities. If it is determined that no competition can possibly exist, then perhaps these properties should be considered communal, just like sidewalks and streets.

The one institution we can establish and enforce fair hiring guidelines for is the government. I have no issue with holding them responsible to hiring cross sections of the population to reflect diversity and fairness. They are the one business we can set standards for, since we all own them.

These moralities that interact with business should be chosen rather than enforced. We can build a society that expects and encourages racial and gender equality, among other basic ideals, without petitioning the government to enforce hopelessly unenforceable bureaucratic sanctions on businesses. Children would also learn a valuable life lesson from this capability for choice; you can choose moralities, but society will alienate and despise you if you choose to hold on to certain obviously uncivilized behaviors. It would force them to define their ethics more intelligently and encourage the cause of unity and civility. Unjustifiable prejudice is firstly harmful to society due to its hindrance of progress, before secondly being unethical and immoral. I want to see racism and sexism completely eliminated, even to an extreme degree, but these discrimination laws do more to hurt this cause than to help it. By giving a person a crutch they don't need, you are patronizing them, and should they accept this unnecessary support, you cripple them. It's time to let society prove its capacity for civilized behavior without the use of force. Otherwise it can never be fully substantiated or appreciated.

The extent to which we should be free is the extent to which we can be allowed to be free and still achieve justice, safety from the actions of others, and the building of a successful society. It should be pursued to its furthest reasonable extreme. Just like the other three governmental functions, it should maximized, but never intrude unduly on the other considerations. The protection of a person's safety and rights is a consideration that freedom cannot cross. So long as we are not allowed to harm each other, safety is being observed. To the extent we are trusted to make our own decisions determines the extent of our freedom. To sell this extent short is to deny humanity's potential as well as your own.

The Bill of Rights Part 2

"Think for yourself and question authority."—Timothy Leary

Our constitution not only establishes the foundation of our government, but also provides us with certain rights that are guaranteed at the highest level of law. It lists actions for which we cannot be arrested or punished. This creates a level of individual responsibility. It is for you to decide what you will do with these rights. Value your right to choice. It can be taken away.

It would be of use to strengthen our federal laws. Expanding the federal law to cover much of what is now being left for states to decide would yield greater benefit. It is ridiculous that a thing that should be called a universal right or crime is criminalized in one locale, but then, taking a single step across the border of a state (that could not function without the strength of the whole), this same thing is legally sanctioned behavior. A child can see the flaw of logic in a system attempting to support such inherent inefficiencies. We have to condition ourselves in order to understand and accept the concept. It has nothing to do with the action being considered socially acceptable or not, and subverts the cause of justice. It is tradition for the sake of tradition, a tradition of boundaries and borders established for reasons only dedicated historians can recall. We must better define the difference between local matters, national matters, and global matters. Allowing for such complications as improper spheres of control create is far from being ideal since, one, the law must never be made into a mockery (especially through a lack of effort), and, two, this situation presents an undue hardship to the police forces of the nation. Nothing quite as terrible as a burden, but they will never be as efficient as they could otherwise be so long as matters remain as they are. Criminals can and do take advantage of jurisdictional and geographic legal boundaries. The police would function more efficiently if they worked as a constant team nationwide instead of teaming up only when necessary. They could all perform their jobs better if the laws reflected their primary objectives clearly instead of creating a tangled bureaucracy for them to enforce, often subjectively, out of necessity.

If we are able to seize efficiency, why deny it for the reason of tradition? To what extent do we hinder ourselves in its name? We call ourselves united; let's prove it. Do we really need to invent ways to vary in our most basic ideals to support our competitive natures? I see the benefits of local governance and the need for the delegation of civic authority to various regional levels. I do not see the benefit in unnecessary complexity. The states are structured differently in several elementary aspects. More should be done to find the most efficient and effective methods of governance and organization, and to then establish that which can be agreed upon and proven as most beneficial. We have the potential to improve the system, but we must make the attempts before progress can be realized.

By what right does the government dare to declare gambling illegal except when they are running the game? By what logic do they make prostitution illegal, but fornication and gift-giving legal, and then expect officers to enforce the law effectively? What fantasy world do they live in where they have the power to efficiently dictate what a person can or cannot ingest or otherwise put into their bodies? Is it all really to save us from ourselves? We don't need them to save us from the thousands of poisons we encounter every day. We need them to save us from murderers, rapists, terrorists, corruption, thieves, war, and destruction. It is obvious that there are ulterior motives at work. They have no legitimate right to make a drug illegal, and it goes without saying that they have no right to, in turn, make relatively more dangerous drugs legal, in the name of public demand. Their true motives are muddled here, and this makes it corruption and a disgrace. This is a petty tyranny, but a tyranny nonetheless. Does the government prevent crime or enforce justice by punishing the self-destructive? Locking crack addicts in cells and forcing the public to feed them is like executing someone for attempted suicide. We are not being robbed of things we should all be doing. We are being robbed of the right to form our own basic moral choices. This is not conducive toward creating a society of responsible free citizens.

The manner in which the government collects its fees for services due, and the manner in which they provide them, are both flawed and in need of refinement. The methods used for tax collection are inefficient, inequitable, and wasteful. This last point alone makes it criminal. Instead of using the most efficient method, or the fairest method, they instead, use every method, and despite creating a tax code of such maddening complexity that few can fully grasp its many ins and outs, some pay dearly while others get breaks they should not, in fairness, receive. Some get trampled through the process, and with such detailed rules, nearly anyone could be made into a criminal. One mistake could theoretically send you into never-ending, perpetual debt.

These sorts of double standards can be ignored, but why, with all our capability, do we not fix the things that are so clearly broken? We need to rethink much of what we now accept as inevitabilities. Why do we not use the incredible gifts of life and capability to improve our system, by a scientific method if need be, always striving for betterment? Why let a tradition of mediocrity continue to hold back the progress that is ours for the taking? We could be proving that freedom works instead of just talking about it. Would this not justify greater respect for the system? Would this not be more admirable and just? For there to be any gain, there must be a concerted effort.

Ideologies tend to call for changes to rights in order to achieve some gain. Communism, for instance, looks to abolish a large number of rights, including the right to private property, inheritance, and wealth. Anarchy looks to abolish various degrees of structure and authority. We should be looking to abolish tyrannies and excessive controls, not rights. Regardless of their extent, tyrannies and excessive controls can only have a detrimental effect on society.

Here is a list of the additional rights that would be of use to society:

- Right to gamble
- Right to equal treatment under the law regardless of sexual orientation
- Right to drug sale and use
- Right to prostitution
- Right to die
- Right to abortion

Establishing and defining morality is an individual's and society's prerogative. The refining efforts of the system should be to eliminate hypocrisy and establish efficiency. These are not the only changes in need of consideration by society. Any breach of justice or efficiency should be considered a high priority by both the system and the people it serves.

Keep the meaning and intention of the Ninth Amendment ("The enumeration in the Constitution of certain rights shall not be construed to deny or disparage others retained by the people.") in mind as I try to justify these proposals. Here are the general outlines and justifications for these rights that I believe should be sanctioned by specific amendments to the Constitution, either in their turn or grouped together as a second bill of rights:

The Right to Gamble

No level or branch of government may declare gambling, in itself, to be an illegal activity.

No one should be allowed to restrict gambling in any way. What is the stock market but a financial gamble that works to support the economy? The same applies to traditional gambling. Saying it can lead to problems is no legitimate reason to justify this subversion of freedom. Eating can lead to problems too. That doesn't mean it needs to be overregulated or monopolized by the government. Life is filled with calculated risks. There is no sense in allowing a financial risk in one place, but not in another, regardless of how the risk is perceived. There is certainly no sense in making it illegal for everyone except for the state. We have antitrust laws that attempt to block such monopolies. Why, at the same time, would we make gambling illegal with the exception of a single party? This creates an obvious coercive monopoly supporting the entity enforcing and benefiting from it.

You could argue that this contradiction of governmental objectives amounts to criminal hypocrisy on their part. Why allow such a possibility when it can be avoided by establishing the freedom to make our own decisions, by acknowledging the potential responsibility that exists regardless of nations? How would people react if a state attempted to make banking illegal except through state-owned institutions? The postal system could be considered a government-coerced monopoly, as well as many states' government-monopolized alcohol retail establishments. Ending these monopolies could possibly be of use if they are hindering society's progress financially or otherwise.

But there is at least something defensible about these services being solely in the hands of the government. There is a conceivably valid concern in regulating drug sales and attempting to control communication to some (ideally minimal) extent. However, there is far less justification for breaking up poker games. When does a vice become harmful or harmless? Is it ever justifiable for the government to intervene where consensual, nondeceptive financial activities are involved? This hypocrisy needs to end, and the appropriate direction in this case obviously is more freedom rather than less. There is no value in creating a citizenry of moral dependents in the name of stability. Local representation is needed across the globe, and state rights are all well and good until they violate basic freedoms or create hypocritical, unjustifiable regulations. Rule of any sort should never be allowed to interfere with progress or justice, especially with no better justification than for the sake of tradition.

The Right to Equal Treatment under the Law Regardless of Sexual Orientation
To ensure equal treatment under law, marriage contracts shall not be limited by the genders of their participants.

Being heterosexual should not entitle you to special rights. Equal representation under the law is a key element of justice. Glaring contradictions are as embarrassing today as past examples, such as, "all men are created equal" except for slaves, minorities, landowners, eccentrics, etc. It is bad enough that prejudiced elements of humanity drove someone like Alan Turing (a chief contributor to the field of computer science and valuable code breaker for the Allied forces during World War II) to suicide for being a homosexual without failing to learn from such obvious, inhuman mistakes. Thankfully, we are not at such a legal stage in the US, but why allow even a shred of such contradictory measures to exist?

Marriage should be seen in the eyes of the government strictly as a legal matter. Again, morality should not be legislated. A man and woman can get married, write their own vows, and in them state to be as promiscuous as they wish with as many different people as they choose. Hypothetically, this couple could continue on such a path, become completely irresponsible in their sexual practices by using no protection, and end up contracting and passing around a sexually transmitted disease through a careless lack of restraint. This would all be more or less legal if all parties involved were consenting to the practices, and they would have no one to blame except for themselves if severe consequences were to arise. Again, it's nice to have responsibility in life. Certainly, it would be a better world if all diseases were easily prevented and treatable, but since this isn't the current situation, we all have to deal with life on its terms.

The point is that if a traditional couple can do this, then how can anyone say gay marriage should be prohibited in the name of decency? To do this, heterosexual marriage would need to be regulated in the name of decency as well. A gay couple could wed and be very responsible in their sexual practices, which (so long as they're legal activities) are really none of anyone's business in the first place. To say that you don't want them to defile the sacred institution of marriage certification is like saying you don't want an argumentative speaker to defile the sacred institution of agreeable discourse or, more to the point, for a secular author to corrupt the sanctity of copyright certification. The government has no place defining or enforcing what is or is not sacred any more than a preacher has any business administering justice. The nation has gay teachers, doctors, soldiers, lawyers, and politicians. To say they can hold public office, but can't marry would be difficult to rationalize to an outsider. Our initial claim to freedom began with the Declaration of Independence. In it, we are all said to have equality and the right to pursue life, liberty, and happiness. To ignore the rights of social minorities, you are putting yourself on the same level as a racist. There is enough hypocrisy already apparent in our system and its history without our having to ignore these principles that have given us so much in return. The Founding Fathers believed in an extreme separation of church and state. We should observe the value in this type of structure as opposed to the one that produced, in the name of sacredness, such atrocities as the Crusades, the Inquisition, and the kinds of extreme, inhuman executions our constitution forbids.

The Right to Drug Sale and Use

No law shall limit an individual as to what they may ingest, inhale, inject, or otherwise consume.

Consider that if a drug is obscure enough, it may be legal regardless of its health effects. Then let's say its popularity rises, and the government criminalizes its possession. Now let's say its popularity were to continue to grow, despite its criminalization, and eventually reach a point where the government could no longer effectively enforce its laws regarding the drug. Eventually the law would be removed, not due to health concerns or citizens' protection, but due simply to public demand.

Is the reason you refrain from using heroin due to its illegality? If it were made legal, would you rush out and try it? If you had to choose on the one hand, say, ten years in prison, while on the other hand, possibly losing your job, family, home, all of your possessions, as well as your mental and physical health, which would you choose? The government should be honest about the facts. The facts are that they cannot effectively stop you from using drugs short of putting a camera in every home, and even then it would not be foolproof. Drug use is inevitable. Of course, murder is also inevitable, but should the government go to such lengths as it does to prevent what an individual willingly does considering their own body? Do you want them locking up people who will do no harm to others, potentially imprisoning a senior citizen for taking what the government considers to be the wrong medication? Do you want your tax dollars to be used to imprison and feed a crack addict who sits in a room watching television all day? Do you want to be told that it's all right for you to drink yourself to death so long as you do it at home, but it's not all right for you to use drugs that are less lethal? If I take a substance that is going to inevitably harm me, who suffers the most from my action? Should we lock up all people who hurt themselves, intentionally or otherwise? If they are harming themselves without drugs, but recklessly, to a point bordering on suicide, should this be considered a crime or a mental health issue? If so, we must imprison mentally ill people who have only harmed themselves as if they were murderers or thieves.

We need a society that holds individuals responsible for their own lives and allows life to teach its own lessons instead of using restrictions to help create a form of mindlessness in order to prevent another form of mindlessness. I am free to take several intoxicating substances that are currently legal. I am also free to buy hemlock plants, which are highly toxic. Imagine that aliens land in your backyard, and they talk to you before anyone else on Earth. They want to know about your society. Imagine eventually having to explain to them how it's legal to buy plants that are highly poisonous, some of which can get you basically as high as possible (like belladonna, for instance, though it does so by taking you inches away from death. Also, it can cause permanent brain damage, and overdosing is very easy), but buying some plants that are much less toxic, and even some that cannot get you high at all, is illegal. It is an embarrassment that our justice system has failed so miserably as to send people to prison, some for much longer than twenty years, over the simple possession of a drug, while there are murderers who have served less than five years before being released and then continued to cause great harm. These are far from being trivial injustices. They are not understandable imperfections. To add to the absurdity of the situation, these people, who the taxpayer spends a fortune to confine, sit in prison being denied liberty at the expense of responsible individuals. All the while, television advertisements promise the solutions to all life's ails can be found in legal drugs. These are followed by more ads selling lawsuits against companies who made the mistake of selling harmful wonder drugs. If they are such an authority on harmful and harmless drugs, where was the FDA while these dangerous chemicals were being peddled? Why not sue both those who are supposedly responsible for these concerns as well as the manufacturers? In such an overly litigious society, it would at least seem feasible. How can we reasonably affirm the government's right to allow us to destroy our health and our minds with one substance as opposed to another? The health risks associated with tobacco make an argument for safety difficult to substantiate. However, even more important than this right is the importance of ending the hypocritical political corruption that supports such unnecessary, inefficient, and damaging control.

In addition to the betterment of freedom, legalizing drugs would cause a massive boost to our economy. The people using these illegal recreational drugs aren't paying taxes on them, and society is paying millions to capture, judge, and jail them. How is any of this fair to people who use alcohol responsibly? These economic gains would be of great use, in addition to the benefits to personal freedom and responsibility.

The government is not protecting us by making drugs illegal. Our justice system's resources would be far better spent dealing with violent crime than arbitrarily enforcing these so-called moralities. These unjust measures and inefficient use of resources must be brought to an end, not only for us to achieve any further degree of self-rule, but also to simply substantiate the self-rule we currently possess. In short, they have no justifiable right to forbid the use of poisons if they are going to sell us poisons.

The Right to Prostitution
No law shall criminalize the act of prostitution.

The question should not be phrased as "Why isn't it legal to sell sex?" The question should be phrased "What has caused it to become illegal?" It may sound like semantics, but what is this so-called moral agenda to criminalize something so widespread, inevitable, voluntary, relatively harmless, and practically impossible to monitor? Is it that unfashionable? It is said to be a crime against public order and is commonly viewed as a misdemeanor. Most practitioners and facilitators simply operate under the radar. They walk a fine line between letting you know they exist and that they are ready to provide the service, but on the other hand, they cannot advertise too prolifically without drawing unwanted attention. Again, we have a law on the books that is being enforced subjectively. This violates the standards by which a law should be made in the first place. Either it should be pursued vigorously or not at all.

It should be noted that this amendment would not prevent states and counties from establishing guidelines concerning prostitution, such as banning street-peddling methods. These businesses could still be influenced to advertise their services discreetly and with respect to the wishes of the community. It could also be regulated more efficiently by establishing standard practices in the interest of public health.

The part that needs correction is not society's generally negative view of the act. The problem is, it exists in every state, and though it is illegal, the efforts against it are more for show than anything else. This is hypocrisy, which is unfair to the taxpayer. Prostitutes, the country over, do not have to pay federal income tax (with the notable exception of certain counties in a single state). They get a free ride by gaining benefits from the contributions of others. How is this fair to the taxpayers? How is it fair to forbid them from making their own contributions to society and pay their own way? Consider that some operators may wish to keep this activity illegal simply because, as such, they can avoid taxation, abuse their workers undetected by any regulatory systems, and then, to add injury to injury, it is often the worker alone who faces the penalties for the act itself. Communities could work together with the businesses providing this service and see each side's concerns best accommodated through the power of cooperation, rather than the bureaucratic nonsense of division and the naïveté of those who think they can control the particulars of consensual sexual relationships. What goes on behind closed doors between willing adult participants is not something we should seek to control. Any effort to do so is a waste of valuable resources. A court of law is a justifiably sacred institution to any citizen who values justice, order, and safety. To potentially force this honorable system into proving if a sexual act and a subsequent transaction of currency was a gift or a payment is a disgrace.

The Right to Die

The right to die with assisted suicide shall be granted in all areas of the nation, subjected to federal regulations, and is to be maintained by the oversight of local governments and medical professionals.

To facilitate the right to voluntary assisted suicide, procedures and standards must be developed, maintained, and continuously improved. Licensed medical experts, legal professionals, and investigators are the ideal facilitators and moderators of these services. The service must be provided in as safely, efficiently, and effectively a manner as possible. To achieve this end, none of the professionals involved in the process may be charged with a crime if they are performing their stated functions correctly.

We are born with the natural right to end our lives. If, for whatever reason, I want assistance in getting it done properly, it is currently considered illegal for anyone to help me. It makes little sense to execute people who want to live while at the same time preventing a person who wants to die from doing so in as humane a manner as possible. Wouldn't it be nice to know your government recognizes this basic freedom? No one wants to think about this decision until the time arrives. Personally, I want to live for every second I can, even if it means struggling to do so, but I know I would rather have the choice than not, and for the government to respect my decisions regarding my own life. Again, it is hypocritical to force medical professionals all over the country to prescribe morphine to dying patents, knowing that if the dosage increases or sustains, the patient will die from it in their greatly weakened conditions. The medical professional is relieving suffering, as is their occupation. To not give these people the protection they deserve is a failure of the system. They should be freed from being forced into a situation where they could theoretically be wrong regardless of their decision in the interest of doing everything possible to relieve suffering, but sustain life. The patient should be allowed to decide their fate whenever possible. A part of the

Hippocratic Oath is to not play god. A god would decide who lives and who dies. A person may choose their entire life whether to end their time here, and suicide has no legal punishment. The government should not attempt to meddle in concerns that should be seen as medical concerns pertaining to personally moral questions. If we can be punished with death, then it should not be denied to those who seek it. We cannot fully grasp their unique situations. We should not pretend that we do.

There should be a strict process set up to monitor this choice. It should be carefully guarded against misuse. Trying to use this legally sanctioned process to kill an unwilling person should be seen in the eyes of the law as attempted murder. This places added responsibility on the citizens involved. This alone makes it a valuable change, not to mention the added benefit it will give our society. It would be an improvement to civility to respect the wishes of our fellow humans.

The procedures should consist of psychiatric reviews to ensure mental parameters (this would also help identify cases involving those who require attention of a different nature), a physical review, a personal investigation to protect from ulterior motives, and a review of the case by a small group of legal and medical experts who are uninvolved with the hands-on investigation. Finally, and obviously, the procedure should be carried out professionally with respect for the wishes of the individual concerned. All of this would also need to be prioritized for cases requesting and requiring greater expedience.

Right to Abortion

No law shall be made limiting a person's right to willingly choose their own health concerns. This includes the termination of pregnancies.

As it stands in the United States, the majority of the particulars regarding the legality of abortion are left to the discretion of state law. State authority is valid in many circumstances, as there is a time and place for localized rule. This, in my opinion, is obviously not one of these areas. This is ideally, like many of the other issues I've mentioned, not a determination for locales, districts, states, or even regions. We need a national, and ideally global, determination on this issue because it concerns basic rights. Even if we are to consider the rights of the unborn as being paramount to the rights of the parent, it would not change the need for a constitutional ruling on the matter as opposed to guidelines that pit state against state on such an issue. The tangles are inefficiencies, and the guidelines are being circumvented by our ability to travel. Critical matters of this kind should not be bound by geographical jurisdictions.

In my opinion, once again we have an opportunity to leave the decision to the individual. Pregnancy should be viewed strictly as a matter of health in the eyes of the government, and as such, it should be left to the individual to decide the particulars of the situation. A fetus is not a citizen until it is born. Pregnancy terminations should be allowed at any point without fear of prosecution to the party in question or to the licensed practitioners performing the termination procedures. Once the child is born, whether by natural means or by any other procedure *that attempts to bring the child into the world alive,* the child should then gain full protection under the law as a citizen of the land. Reproduction should be a selective choice. To allow the government to influence this choice is a form of population control. This is especially relevant if we ever face global overpopulation issues, as the government may then be inadvertently forcing a child to be born so that it can starve to death in the name of humanity or morality.

This line is the most difficult to draw of all the potential changes I've proposed. On one hand, you prevent and, indeed, take the life of a child; yet on the other hand, abortion is as inevitable as rain and, in some circumstances, necessary for the mother to survive. In this way we are all in favor of the right to abortion to some degree. If there are to be any exceptions of any kind toward forbidding the practice, such as problem pregnancies or rape cases, then it should be up to the mother to make the determination and not governmental guidelines. There is enough bureaucracy already without having to add more. Again, it is a moral question and should be left to the individual to decide. Should all women who've had abortions be locked away or executed for first-degree murder? Should the doctors who have performed these procedures be considered mass murderers? If a fetus has rights, then at what stage of its development does it gain them? Is it a man's legal responsibility to save the lives of as many of his sperm as possible by putting effort into seeing as many as possible turned into living persons? Without use they die, whether inside or outside of the body. Should every single menstrual cycle ideally bring forth life? Should vasectomies and tubal ligations be considered murder as well? You are, after all, killing living tissue, which, if left unaffected, could potentially bring forth life.

Abortion, even when illegal, has always been available to women either through procedures given alternate names or performed illegally, where the mother's safety is embarrassingly disregarded by the system. Instead of having laws that are ignored in specific circumstances, the system should accept abortion as the inevitability that it is and deal with it maturely.

Those who protest this right should spend more time establishing programs that encourage expectant mothers to consider giving their children life and putting them up for adoption as opposed to aborting them. Instead they attempt to use the government as a means for imposing their morals on others. If they really cared, they would take some positive action instead of trying to criminalize a massive cross section of society. These women come from a myriad of different circumstances. I admire those who choose life in these difficult situations. Perhaps those who deserve the most credit are those who stand to gain nothing from the birth process, but go through with it anyway just to facilitate the chance of life for their child. This does not, however, give me the right to say there is never a time for abortion or to establish a moral mandate on the issue. It is not our decision to make for others. It is for the individual to make these biological decisions involving their body and what lives inside of it. We should place no personal judgment on anyone for having chosen this procedure, regardless of their circumstances. It is none of anyone's concern what a person does with their body.

Abortion is birth control, and where you draw the moral line in the reproduction process is up to you, but where the legal line should be drawn is through the most enforceable stance. The coat hanger cannot be monitored as effectively as the process of birth. To protect the unborn equally and by the blind standards of justice could arguably entail criminalizing all forms of birth control. Since the line must be drawn somewhere, the government should stay out of the moral situation as much as possible. This, like the other rights I've proposed as being of use to society, is meant to protect those who, as it stands, are being treated with undue discrimination. One last thought on the subject: if men carried babies instead of women, abortion would have always been fully legal.

Directions Please?

Once we determine where we want to go, how do we get there? These are the harder answers to come by. Criticisms abound; solutions are sparse. It's easy to say that unity, among other things, would be an obvious improvement, but making these changes is a different matter entirely.

One way to achieve progress is simply for more and more of us to recognize the value of progress. As the numbers grow of those who value positive ends, the task of Utopia's construction becomes easier and easier. The only way to make these numbers grow is to spread the proper values. Demonstrate to others the power of cooperation. Multiply the desire for knowledge. Identify and communicate the positive aspects of our world for reflection and encouragement. Identify and communicate the negative only to target and study the areas requiring improvement; never brood on them or incite anger in yourself or others.

Never encourage uncivilized or useless, pessimistic attitudes. Only encourage that which takes us closer to Utopia. Never say, when an opportunity for betterment makes itself available, "We aren't ready for this advancement." We are always ready to move closer to our goals. Remain open-minded in your own views, as we can easily make mistakes in reaching what we see as a better world. Many, if not all, of the very worst of mankind's atrocities have been committed with some good in mind. This is a snare we cannot afford to fall into if we are to one day face seemingly insurmountable obstacles.

Just as other nations have adopted the aspects of our governmental system they see as being positive, so should we be looking to other nations to see where they are excelling. Perhaps the countries with the lowest crime rates should be examined to see how their justice systems operate. Maybe their prison systems are functioning much better than our own. Maybe their methods of punishment are more civilized and fair and thus yield better results. There are some prisons in the world where a one-year sentence for a relatively petty crime is, in effect, a death sentence and is especially so if the accused happens to be physically weak. This is not conducive to the cause of justice. It is a poorly functioning rehabilitation system that encourages destructive mentalities in those it seeks to reform. These shortcomings cause severe backlash against society and should be given much greater attention than they now receive. To deny the possibility that we've missed something in our hastily constructed, yet ever refining structure would be arrogant and unrealistic.

Here are some ideas on how to achieve the seemingly unachievable, but first a note on power.

Power is not found in pointing a weapon at someone and forcing them to do what you want. Anyone can force results with this method. Real power is in reaching the objective without the gun. It is done without threat or coercion of any kind. This is power. The art of persuasion can be used to deceive, or it can be used to encourage positive outcomes. It is more powerful than any weapon, because behind every intentionally successful use of a weapon was a person motivated enough to employ it. We must never resort to aimless abject violence to solve problems. The reason for mentioning this is that we must be on the lookout for abuses of power and take action against these setbacks. Further criminalization of political corruption would be advantageous to our goals.

Civility

To achieve civility, and ultimately unity, we must promote them en masse. Through actions and words, express your desire to live in a world of cooperation. Encourage your representatives to state unity as a primary concern. There is little point in trying to establish these ultimate goals with force. A degree of civility is created by the world's justice systems, but trying to make unity a mandate would be like trying to enforce economic efficiency and success. A combination of activism and pacifism will be needed to counter the effects of uncivilized actions and agendas. The ideal combination will allow society, as well as the government, to function to their greatest extents. We can be civilized and still stand up for what we believe. We can seek peace without abandoning our principles, rights, or the cause of justice. When you most desire retribution, consider the situation carefully. Never take the advantages of a civilized society for granted. Continuously consider what brought about these advantages that surround us daily. Spread the admiration you find in others. People who have faced the difficult task of promoting civility in barbaric times and environments of little visible hope deserve much of the credit for the civility we currently share. They should serve as role models for the rest of us. Civility is easier to establish during peacetime than in war, but it is perhaps recognized and appreciated to a greater extent when contrasted against the obstructive shroud of conflict. Never underestimate the influence that spreading this crucial ideal can have. Ask yourself what kind of world you want to live in as you age, and seek to nurture that vision. It is not only the global efforts that will bring about massed unification. It will require efforts in our daily lives as well. If we are to achieve the pinnacle of human achievement, this is unquestionably one of our greatest current obstacles. It is not so much that we can all make a difference in this goal, it is more accurate to say that our combined efforts are absolutely required if the benefits of peace and cooperation are to be fully realized.

Consider those who have died in the effort to establish safety for us. Consider their contribution to the causes of freedom and prosperity. What is the correct way for us to honor their memory? Should we disregard the cause to which they paid the ultimate price? If any degree of safety is the concern, whether it is individual, national, or global, then what can possibly establish it better than civility and unity? Most would answer that we should support this effort, but understand that supporting it may involve us setting aside trivial concerns to prioritize the greater need. It may require people locked in perpetual vendettas to set aside comparatively irrelevant concerns in the face of a more valuable end. It may require a great deal of consideration for the concerns of others.

Safety

It is important to consider that civic crime tends to flourish to the greatest extent among those who have the least hope in life. Helping them gain a reason to live, and encouraging them to hope for a better tomorrow, would be of great use in eliminating crime. As previously stated, civility and unity are our greatest weapons in the fight for safety. The worldwide desire for attaining civility's ultimate form is the surest way to obtain security from harm on a national level. We should be eager to find peaceful solutions where the safety of future generations is at stake. Were the responsible parties in the areas of the world's greatest conflicts to place their grandchildren's safety and well-being above lesser concerns, they would not only encourage betterment, but would also possess greater honor. Any coward can incite conflict and stir up emotions of hatred in others by illustrating in detail (through exaggeration and overgeneralization if need be) how they are being oppressed, by suggesting that miseries are being mocked, by pointing solely to the negatives, and by convincing people that the destruction of the perpetrators of their ails is not only just, but that it is the only reasonable option.

When violence is called for in defense of a society, this message will not need to be preached to such an extent that motives wear thin. Political institutions proclaiming, in so many words, "Just take our word for it" is not credible enough either. Providing the facts of the situation should be enough to convince the public of the necessity for violence if it is indeed needed to ensure safety. The cornerstone of democracy is the assertion that the judgment of the masses has, to a degree, a greater value than individual opinions. When the will of the people is ignored by their representatives, it is a contradiction to call the political system a democracy or a government of the people. If we desire civility and unity as a majority, the desire for safety and the deliberation of the people and their representatives should be enough to drive any necessary enforcement of safety, unless, of course, their representatives have ulterior motives. If corruption is proven, the aforementioned strengthening of laws against such activity would be called for, since inciting an unnecessary war and thereby causing murder on a grand scale should be considered the utmost of all crimes against humanity. One of the greatest proofs of our species' incredible potential is that there is no short supply of people among us who are ready and willing to risk their lives to defend society. These people are mankind's greatest resource.

Prosperity

How do we achieve the greatest prosperity? Achieving the other factors will encourage and facilitate it, although economics is the specific field of study that attempts to find the best route to economic success. Since our overall knowledge is growing, our chances of finding, retaining, and refining the optimum route are improving. There are clues to be found in the past. Using evidence and logic with due consideration for all goals, sound economic philosophies must be developed and put to use. They are almost always defined by the extent of the government's involvement in economic issues, and as such are within the sphere of political philosophy. Society's efforts for prosperity must be seen as working hard to achieve it, while being freed from influences such as detrimental government actions. Beyond this, society must see prosperity as a worldwide goal and establish it in a creative fashion as opposed to throwing money at problems without due concern for its use, which is, at best, a temporary solution and, at worst, a harm to independently self-sustaining prosperity.

Economic prosperity will be gained with greater ease if we can achieve greater agricultural efficiency, technological maximization, and so on. The ideal society is not only prosperous economically. It is also prosperous in production and in providing the goods and services people both require and desire. We must employ highly detailed observation, thorough documentation, and careful study of each step we take. Each must be determined as steps bringing us closer to or further away from our ultimate objective of progression.

Knowledge

Knowledge is a vital element of Utopia's foundation. Without it, the ideal state could not exist. We must agree that knowledge is of value, that it saves lives, expands our understanding, and makes life more interesting, fulfilling, and comfortable in order to proceed to other challenges. We need to agree that knowledge should not be restricted to certain ideas, but that it takes all aspects into consideration. Opposing viewpoints and even obviously flawed assertions must be studied thoroughly to understand the ideal paths more clearly.

Book burnings are the loudest cries of mental weakness imaginable. They are remnants of a repressive era. An idea is only exceptionally dangerous to people who are feebleminded enough to be irrationally and blindly swayed by it, and the orchestrators of book burnings fit an old saying exceptionally well: "A thief thinks everyone else steals." They assume everyone is easily swayed because they have been easily swayed themselves.

The smaller version of banning or destroying books is censorship and needs to disappear as well. You can watch a movie filled with censored words, and many a small child could fill in the silenced portions if quizzed to do so. Attempting to protect them from forbidden combinations of syllables is unnecessarily condescending. The people trying to eliminate words are the ones who give them their mystique and their forbidden nature. A more sensible approach would be to place more concern on implications and less on specific words. Anyone can form a terribly profane, hateful, or disgustingly immoral sentence without using or even hinting at these certain forbidden words. Anyone could write a collection of sentences so vile and coercive of explicate, degenerate, criminal behavior that even under freedom of speech they could possibly be arrested, and all of this without using a single one of these dirty words we all know so well. Children are first mystified by this social phenomenon. Then, once they learn the harmlessness of the words, they see adults as attempting to hide harmless things from them, and all because they feel children are incapable of restraining themselves from compulsive imitation of everything they see and hear. If shielded from harmlessness long enough, the children may even begin to believe it themselves. They should be taught to control their language, since it is useful to the skill of selective communication. However, they should not be taught to censor themselves exclusively from specific words, but rather from carelessly using expletives and slurs. Compulsively censoring your own language with "less offensive" alternative words is more offensive to some people than if you had just used the big bad word, since it comes to mind either way and treating others as if a word can harm them can be insulting. Variety in language is appealing. We should all learn how to communicate appropriately in regard to our given situation. Other cultures are accustomed to a different set of communication standards than our own. How can we hope to communicate effectively with them and adapt to an ever-uniting world if we restrict ourselves to a single mode of discourse to suit our personal taste or comfort? In short, grow up. It's just sound; it can't hurt you. Adaptability is a positive trait, not a negative, and more focus should be paid to implication and intention, and less to individual words.

Knowledge can be obtained by instilling a love of it in people. We should all be encouraged to study as many subjects as we can and to the greatest extent we can manage. The sooner we develop a taste for it, the more we can accomplish in our lifetimes. The first years of a child's education should be spent simply introducing them to the wonders of our incredible world, not attempting to prepare them for a life of turmoil so they can simply survive. It is no small wonder that if the latter approach is taken, few can find the motivation to appreciate the subjects of study. Education is the key to raising knowledge in society. Schools, colleges, museums, and libraries must be improved to help raise the potentiality for knowledge. We need teachers who are committed to teaching instead of going through motions. We need facilities that excite the desire to dig for information and understanding. We need parents and guardians who share their own wonder with those they are held responsible for guiding. In order to do this, the parents and guardians must have subjects they find interesting themselves. The desire to learn can be, and ideally should be, contagious. This may help the cause at hand more so than anything else. The wonder of our situation is not something everyone appreciates. You have to find something you enjoy learning about and, starting there, move on to tackling subjects you dislike. Seek to unravel the aspects that are mysteries to you. To stimulate wonder in children, it may be helpful to present them with unsolved problems so that they can gain an understanding of how deep the waters are and to know what they are up against if they plan to pursue a particular field to a greater extent (choosing a career).

As I mentioned in the introduction, I am attempting to limit the scope of this writing as much as possible to societal matters. Keeping that in mind, education has been considered an imperative of the government for much of our past (in ancient cultures as well as our own). Many have suggested that the government should not be involved in education and some vice versa. If it is to fall into one of the four previously mentioned governmental imperatives, it would have to be that of general welfare, though it should be noted again that the government is *providing* it to a large extent as opposed to only *promoting* it. It could be that schools would function better solely as either society's responsibility or as the government's. Though a combination of efforts may also prove to be the most ideal approach. Regardless of whichever method proves to be most effective and efficient, the standards should be continually raised. Methods need to be reevaluated constantly to determine where we can make practical improvements. Exposing students to our professional sectors to a greater degree may be worth exploring so as to provide a more well-rounded education. It could also be that critical thinking is being significantly underrepresented in the face of redundant information retention. The practical concerns are not the sole concerns, and the ability to theorize and think creatively must not be shortchanged in the balance of a well-rounded education.

Having students spend more time in libraries, zoos, and museums may also be a good idea. I am not a professional educator. Those who are must be creative. They must collaborate and keep their ultimate objectives absolutely clear. Mistakes and accomplishments should be noted and studied to find opportunities. As there will always be room for improvement and efficiency, the educational systems should constantly seek to refine their tactics and strategies for better encouraging enlightenment and achieving greater knowledge.

Health

Again, this book is not meant to outline my political philosophy. My focus is social betterment, and I only touch on political barriers because they hinder society's progress. As I've pointed out, there are several responsibilities I would like to see the government hand over to society. There is, however, another modification I would like to consider that I think would yield benefit to our social structure. Once again it involves the government, and once again it is a touchy subject. This time, instead of taking a responsibility away from the government, it may be beneficial for them to take one (*only in part*) away from society. Let's keep in mind the four things the government is supposed to concern itself with as I try to justify my proposal.

Before I begin, I would like to note that currently many of the social programs in the US are obviously flawed, and many could use reevaluation or even termination. Adding more of them may be a very bad idea, yet…what is the Supplemental Nutrition Assistance Program (the food stamp program) but a limited form of socialized health care that focuses solely on the threat of malnutrition among the poor? What would legally mandated health insurance be but a socialized health care system with a middleman? If the government sets restrictions on what insurance companies (auto, health, or otherwise) can and cannot charge or practice, then has insurance not become a part of the governmental system? If not, they might as well be, since the government forces its citizens to accept certain services they provide as being just as mandatory as are taxes. In many situations, health insurance is a practical necessity to obtaining decent health care. This service's takeover is a convenient abandonment of capitalism by the system. It is very convenient if you happen to be an insurance company or health care provider since people will be forced to buy your services. Part of what causes this is the extraordinary costs associated with health care, which we are all determined to need sooner or later. We can reduce these costs to a reasonable level by cutting out the middleman, not by giving the middleman advantages or by adding more of them. Hospitals are able to charge their grossly inflated prices because they deal with insurance companies rather than individuals. If they dealt more with individuals and let competition influence the fees, they could only charge reasonable amounts. As it is, the higher the number of insured individuals, the higher the amount hospitals can charge. High amounts would be fine, since we all want the best health care possible, but amounts that only a small percentage of the population can afford makes health care into a kind of inefficient drain on resources. A public health care sector would help to get these costs under control and would eliminate the need for a great deal of government involvement in the affairs of insurance companies. In this system, we have to start paying insurance companies for our health care costs gradually from the time we get our first job, which incidentally, is the same time that we have to start paying for our mandatory retirement program.

Many argue for and many argue against socialized health care. Again the question revolves around the government's precise responsibilities, which are debatable. The pertinent question in this case being, to what degree is the government responsible for our safety in medical concerns? Do serial killers and viruses not have a similar effect on society? To suggest that the government has a responsibility for this type of safety opens the door to several questions. In my opinion, it could work and could work well to have a public health care system. Most other developed countries have these types of systems in some form. Their various degrees of effectiveness and efficiency are debatable, but as with all other governmental affairs, there are several contributing factors to success and failure. Regardless, one could argue that the United States (and any other developed country) already has socialized health care to an extent, since if a major health crisis, such as a virus, were to begin destroying large numbers of civilian lives, the military would obviously become involved in alleviating the crisis. This would be the government protecting the lives of its citizens in a medical matter. They would protect from major widespread threats, but not from minor, individualized threats. As I've already implied, what would the militaries of the ideal society have to fight except environmental threats? Once the threat of war is brought to a near nonexistent low, there will only be environmental threats and civil threats to reasonably consider. Militaries will eventually have to join in the war on detrimental health concerns to see any lives saved or liberties preserved.

Since I've already said that our health is our own responsibility, it may seem like a double standard to suggest that it is the government's responsibility as well, but we can easily deal with this line if it is drawn carefully. Many would say that if you make drugs legal and then provide social health care, drug addicts would soak up massive resources for which the taxpayers must pay. I agree that this would be unfair to the persons desiring an efficient public health care system. Some have used these kinds of statements in an attempt to shut down the discussion, but a sensible society could solve this problem. Let's say a heroin addict goes to the public system and claims health problems that the doctor then determines are effects from drug use, and gives the patient information on rehabilitation services and treats their health concerns. Then let's assume the same patient comes back a month later with worsening conditions, seeking treatment for a new ailment. If the doctor determines the patient is still causing the problems through their own actions or neglect, the doctor should be required by law to refuse treatment and point the patient to the nearest capable, privately owned health facility. If the patient cannot pay the private sector costs, then they have left the system no recourse but to deny them treatment. If they have the finances to buy drugs, they should be able to buy health care services. This would also need to be applied to tobacco users, people who can't get their diets under control or proper exercise, and so forth. If there is to be a beneficial public sector, it should provide its services with a respect for the wishes of the people paying for them and with both accountability and compassion in mind. If such onsite decisions were required of our hypothetical system, who but a doctor could make them? Certain ground rules such as this would need to be established to provide fair, efficient services to the public, but it could be done.

Another benefit of having a public health care sector is that it would yield greater choice. People could get free health care, which may or may not be as good as the private sector could provide. Ideally, the people who don't mind spending the extra money should be able to go to the private sector to receive a higher value of care. In such hard decisions as our health concerns can be, it would be greatly beneficial to have choice. A person may have a condition and have the extra resources to seek the best possible treatment, or may have others depending on them and choose to pursue the cheaper route in order to save money. In any case this should be their choice and their choice alone. We should seek to have these kinds of options in life. To determine and prioritize your own life objectives is key to true freedom; it is key to being a responsible individual. We are being robbed of our ability to make decisions, plan our futures, and live life to the fullest so long as we have few choices. It would be of use for people to know from birth that they can receive treatment for life's ails so long as they value their health. It would be of some foreseeable use for a person to know that when they get too old to take care of themselves, they have a public nursing home available to assist them, which, granted, may not provide an ideal amount of care, but which would be humane enough to help those who cannot provide for themselves. These concerns must either be addressed by the individual and/or their family, compulsory contribution (the government), or by voluntary contribution (charity). You may decide that you want to plan and save up money for your twilight years so that you can afford to buy the kind of care you want. There is value in having these kinds of choices. Your life should be, at least in large part, your own responsibility. Society and the government should work hand in hand to provide you with options, not restrictions.

The crux of my ideals regarding health care systems has less to do with hospital ownership and more to do with creating a system where health insurance is not the necessity it is in much of the world today. Notice, though, that I am not saying insurance should be abolished. It should be an option for those who see a personal benefit in it, rather than the intrinsic social feature it has become. There are obvious inefficiencies in a system that charges such absurd prices as many medical practices currently charge. The poor, too often, cannot possibly afford basic health care. This makes the middleman mandatory. Regardless of which system you prefer, these grossly inflated prices should be seen as something to eliminate. Knowing more and more that people rely on insurance tells health care providers that they can charge whatever the insurance companies can afford, as opposed to what the patients can afford. So long as everyone is insured, and given that the insurance companies can maintain their costs, the situation can remain as is and function, but what about those who cannot afford the insurance costs? If the government is going to pay their bills, why not pay other taxpayers' bills? These measures, like social security, possess glaring contradictions where some are paying for services that give them no benefit. This could be construed as taxation without representation, which is one of the grievances that caused the Founding Fathers to rebel and form the nation in the first place. Doctors should be paid well for their often necessary services, but their compensations must come from a free system rather than one being dominated by what should be seen as an optional service. Anything that weakens the insurance companies' legal and practical mandate over our lives would be of benefit. We now have a greater capability to enforce litigation than in the past. A court could seize property, revoke a person's right to drive, and so forth to easily eliminate the need for mandatory auto insurance. We can function without the universal belief in a service that is being used to rob us rather than to provide us with a solution to certain valid concerns. Understand, I am not necessarily calling for a socialized/private health care system yet, though I am currently leaning in this direction as an ultimate goal. The issue is better left to detail in a political philosophy I have yet to formulate and where personal, detailed, concrete opinions must be stated. Still, society should have a concern with the issue and carefully consider all aspects.

If we are to socialize health care, we would first need to determine if the government can have any positive effects on public health and also if their efforts are of greater effect than those of the private sector. If they are a total inefficiency, then this should not be considered as an option, and at the most, the government could consider giving vouchers, where absolutely justified, to the poor for use in privately owned hospitals and doctors' offices. The issue should be viewed through many lenses, such as hypothetical extremes. If a deadly epidemic were to break out and conditions reached a catastrophic pitch, it is obvious that the government and the public would both be doing everything possible to address the situation. We would work as a team in such dire circumstances. Would private hospitals play a role in the medical emergency? Would the government be in a position to provide assistance that could be considered unique or more effective in some regard than the efforts of the private medical community? The military may have resources unique from the resources found in private sectors and local governments. Are we handicapping ourselves by allocating wasted resources to government programs on the one hand, while the government neglects areas of potential opportunity on the other?

The only way such a health care system should be considered for adoption is if the myriad of other welfare programs are first either justified or terminated. Food should not, ideally, be a governmental responsibility, neither should shelter. A well-functioning economy and society can support these needs on its own if properly facilitated. Potential threats of a sort best handled by society will always exist, just as some threats are best dealt with by the government. Knowing this, there will always be an opportunity for the government to perform functions society struggles to address and vice versa. As responsibilities are passed back and forth, there will always be overcompensations and undercompensations. A degree of misallocation is inevitable.

The key is for the concerned parties to accept only those functions which they can perform best. Justice is an excellent example of a proper governmental function, though even it must be assisted by society to an appropriate extent to achieve positive results. To establish the hallmarks of the successful society, we must determine what considerations are best addressed by society alone, which are best left to the government, and which require measured teamwork to give the nation, and ultimately the world, the results we all seek.

Throughout the world, responsibility over health lies in both our own hands and in the hands of the medical community to various balances. To best further the cause of societal health, we must facilitate the needs of both parties. As with the other factors, health is interrelated. Technology, knowledge, prosperity, safety, efficiency in dealing with social issues, and the collective will can all play a factor. Unity will also contribute to this cause since medical communities sharing their findings and data more openly and acting as a team would yield greater efficiency and be more apt to produce results. Also, with unity, the masses can approach an obstacle to medical science as a team. Mechanical engineers may be needed to solve certain complex problems relating to surgical technology. Lawyers, philanthropists, organizers, community leaders, and individuals all play important roles in this effort to achieve the greatest possible physical and mental conditioning for society.

Technology

Technology is already making gains at an unprecedented rate. It should be continually pursued, and just as with the other factors, there is always room for improvement, but the methods driving it are currently functioning very well. The greatest aspects hindering technology's advancements are the people who oppose it or those who are reluctant to adopt it. They are like children who complain, "Why do I have to learn how to read?"

Some oppose technologies such as the Internet. The Internet is a communication tool. What these people really oppose is communication, typically because they have no faith in other people's ability to understand and discern. Others oppose advancements they believe will harm society due to their usefulness. If something is invented that adds efficiency, to say we should not employ it because it may cause economic harm is saying we are purposefully inefficient to produce better economic results. This assertion leads some to claim that some of the emerging technologies should not be developed due to the harm it will cause employment levels. Usually this does not mean fewer jobs; it just means different jobs. If this attitude were correct, then we could create enough jobs to employ the entire planet by simply destroying all technology. Look at the jobs we could create by banning bulldozers and heavy equipment. Look at the harmony we could create by enslaving the world. Look at the freedom we could create with anarchy. Look at the peace we could create by destroying the entire human race. This is not the future we want. Some could learn that we've discovered the cure for cancer and still be jaded because some gravediggers might lose their job over this advancement. Accept the challenges and responsibilities our tools and technologies bring. To say we can't cope with them would be on the level of our early ancestors saying, "I can't figure this shovel thing out, so I'm just going to keep digging holes with my bare hands. Besides, if those things catch on, it's going to put some of us out of work, and then where will we be?" Learning new devices may be difficult, but we should value the opportunity to do so.

Independence

It should be noted that there are groups in this country who represent interests diametrically opposed to freedom and choice. We must support our concerns constructively and strategically to combat the lowest and most unsound proposals. Take, for example, certain groups, which will herein go unnamed, that have taken such initiatives as protesting soldiers' funerals, holding rallies for racism and separation, and who take sickeningly degenerate stances in several moral and social issues. Understand that I am purposefully generalizing here, throwing several groups into one classification. Whether they are racially motivated, religiously motivated, ideologically motivated, or whatever else may come along in the future, several groups fall into the category of factions most would readily and justifiably condemn. Many are probably already aware of the best approach to these situations of obvious social hideousness, but I have noticed a need for some clarification on the issue.

How do we view and deal with such radical groups constructively and effectively? Many have suggested and performed counterprotests, but this only plays into their hands. Radicals often understand that the only way they have of gaining recognition for their deeply flawed assertions is to stir up strong emotions in the masses. The thinking being, if you can't convince, offend. This is to compensate for their hopeless lack of clear, demonstrable substantiation. By giving them recognition of any kind gives them a voice they would not otherwise possess. By demonstrating our opposition to their causes, we give them a platform they do not deserve. The key is to break the power of these groups' leaders. To do this, convince their followers to abandon the causes by detailing their lack of value. Attack ideas, not individuals. Have concern for them and treat them as fellow searchers for the ideal, and they may wake up, but never acknowledge their pathetic cries for attention. Rational debate should never be shunned, and people should remain free to follow even the most absurd of propositions, but this does not mean we must acknowledge that which only deserves condemnation or scream in vain into the face of minds that are as closed as coffin lids. Offending them may simply reinforce their positions as they often do our own. My suggestion is to ignore them to the utmost extreme, even to an almost unrealistic, bizarre extent. This is why I refrain from mentioning them. When their signs are being waved and their slogans shouted, the masses should pretend that they do not exist. Give them no media or individual attention whatsoever. A part of what fuels society's anger against them is the way in which Americans are portrayed to the rest of the world by the views and mentalities of these social minorities we could all happily do without. The image would be instantly repaired, at least in part, by a total rejection of the proposals, while still supporting our rights. By broadcasting their slogans, we give them strength. By attacking them with violence or even peaceful scorn, we turn them into martyrs for a taboo cause that may then appeal to rebellious types who are desperately seeking any cause to follow. If you are to attack them at all, do so by questioning their proposals and logic. Ask what good can possibly come from their assertions.

We should also consider that in many countries of the world, these groups would be killed after their first protest or public demonstration. In a strange sort of way, they are demonstrating the extents of our freedom, a comparatively extreme amount we should cherish and never take for granted. One can even smile over the fact that, despite their ideologies being disheartening indicators of humanity's lack of social progress, we can still sleep at night knowing that the secret police aren't going to detain them and subject them to a "political reeducation program" which amounts to inhuman torture. Regardless of their potentially destructive ideologies, they are not targeted for assassination by government agents. On the contrary, they are protected by the system they condemn, a system without which they would be destroyed either by governmental tyranny or by mob violence. Perhaps you can now understand my reluctance to name names. They are not worthy of attention or recognition until they prove that they are seekers, as opposed to knowers who lack decent substantiation. We are almost thankfully faced with too many serious proposals to consider wasting any time on obviously worthless doctrines.

To maximize self-rule and achieve greater freedom, seek to make your own choices and accept the responsibility for your actions. Expect others to accept their own share of proper accountability. If you feel the government has no right to restrict your freedom in some regard, you should continually examine the situation and, if convinced, voice your concerns. Petition the government for your rights as a team. If concerns remain unaddressed, find new representatives who will pursue your objectives. Expect results where results are due. The First Amendment states that we have the right to "petition the government for a redress of grievances." Exercise it creatively. The government does not have to listen to the public, but the representatives do if they seek a political future. See the section on high-level changes below for other ideas concerning democratic efficiencies.

Comfort

Comfort will come naturally as we achieve the other parameters. Prosperity allows for it, but safety, independence, technology, health, and knowledge are also required to establish a life of greater contentment. We must encourage civility among the violent, independence among the subjugated, knowledge among the uneducated, health among the sick, economic prosperity among the poor, and so forth. Pursuit of these factors will allow future generations a higher general degree of comfort throughout their lives. Our goal should be to seek fulfillment and to increase the potential for it globally. In order to do so, fulfillment itself must be well defined. What do you personally need to attain it? What would make you happiest in life? Others may require likewise. Comfort and the potential for fulfillment are gauges and defining characteristics of the ideal state. These indicators are the prize, whereas the other factors are how we can go about achieving it. The more we attain the prize of comfort, the more capable we will become in all aspects and endeavors. One aspect reinforces the other to ever-increasing achievement.

Efficiency in Dealing with Social Issues

To gain efficiency regarding social issues, we need to study the political, judicial, and societal systems of the world. We need to create efficiencies at all levels and explore the aspects of viable political philosophies to spot opportunities and perils. Achievements in administrative efficiency must be acknowledged and duplicated to achieve positive results. Areas of concern must be examined and methods refined. As with the other imperatives, the goals will require massed concentration. Organize, form proposals, and identify proven solutions. The professionals of our most important fields must be looked to for the greatest part of the answers, but that does not mean that laypeople cannot also make their own contributions. Organization is a vital concept when creating efficiencies. First, identify inefficiencies. Next, define them in detail. Then, consider all forms of combating them, and weigh all considerations and possible side effects carefully. Finally, choose the best course of action and implement it. Utopia will know how to best deal with many of our most challenging social issues, because it will have learned from our struggles. Their leaders will have the benefit of hindsight. They will clearly see and understand what has caused our successes and failures.

The Preoccupations of the Collective Will

Unity will allow for these desires to be put into motion. As to what they should be, society must choose its own areas of focus in building a better world and who can best achieve them, whether it be the government or society. Focus on construction more so than demolition. Build beautiful cities and monuments to humanity's worthy features. Refine aesthetics and value humanity's achievements. Build *functional* monuments to education, health, commerce, justice, nature, and human existence. Seek to create the future instead of recreating the past. For the preoccupations of the collective to do anything useful, they must have a small degree of uniformity, enough to value cooperation, though aesthetic matters will require a diversity of opinion to be fully realized. The masses must have access to information regarding society's areas of strength and weakness before they can exercise their will in a meaningful fashion. Consider the beauty created by the sharp contrast between our natural world and the man-made world. Together, our search for beauty can yield results that reflect humanity's incredible achievements.

High-Level Adjustments

To speed Utopia's arrival and to maintain our current progress, we are sure to encounter situations requiring high-level decisions. These are changes to the big picture, such as governmental changes. The government of the United States has a functional foundation, with the three chief branches to keep the whole in check. Any stable government has these power checks in some form or another. They exist to form organization, focus, and minimization of error. Somewhere between the extremes of individual rule and majority rule exists the ideal state where all concerns are best represented, but even Utopia is not free from mistakes. It is just better equipped to both prevent and deal with them.

One problem with our system, regardless of the changes you personally feel need to be made, is found in a somewhat prevalent malaise over our current situation of possessing only two practically accepted political parties. Two agendas yield choice, but not very much of it. Having the highest possible priorities laid out as a mutually exclusive set of decisions is far from ideal. One might argue that if the government of the future is as united as I've suggested it will be, then shouldn't there be fewer decisions since everyone is in basic accord as to how we should best proceed? Shouldn't unity ultimately yield a one-party system? I am not suggesting this uniformity of thought is an ideal destination. People should be able to agree on the basic issues while still allowing their collective views and contradictory opinions to throw light on what current issues are in need of the greatest attention. It might be that the political parties of the future are called by names like (for example's sake) the Education Party, which focuses on the betterment of potential knowledge by channeling tax dollars into the military study of scientific threats and encouraging (to the appropriate degree) the resources and level of attention paid to knowledge and discovery in society. In short, this would be a group whose primary focus is knowledge. There are those who would argue the valid point that if we were to take this single aspect of our society to its greatest extent, then the rest of our issues would fix themselves. Perhaps there would be people who vote strictly for the Education Party for ten years only to see a greater need for the advancement of another aspect and start voting for the Civility Party, who see focusing on creating a more peaceful environment as yielding the most benefit, or the Freedom Party, who believe in maximizing liberty to make the greatest progress. Regardless, this situation would only see fruition if the basic objectives continued to develop and the correct beneficial ends were agreed upon. Obviously any political party in any environment would have to see the big picture and not focus exclusively on a single aspect while ignoring everything else, but it may be of greater benefit for them to have specific agendas as to what are the most beneficial objectives in light of the current situation.

This is not necessarily my ideal view of the future of politics. I don't know what will become the ideal situation regarding political parties. Perhaps a Utopian Party would help us reach a better future simply by centering on improving the basic categories to their utmost degree and focusing more on the long-range goals than the immediate ones. To be of use, the parties should represent the people's desires. The current situation is two groups, each possessing two bulletin boards, one marked as "our achievements," the other labeled "the other party's mistakes." This is inefficient when a person sees certain issues as vital, and is useless when, regardless of their vote, their greatest concerns are nearly guaranteed to go unaddressed. What is a person's motivation to vote if, either way, their chief concerns are sure to be ignored? Perhaps some of the people who have the least motivation to vote possess the highest motivation to seek out a better world, if only they had the opportunity to do so. We must create these opportunities. When election after election comes down to choosing what a large portion of the public sees as the lesser of two evils, there is an obvious problem with our methodology. Having a more influential third party could help, but it could become practically as useless as the process is now. Many other countries possess several parties, and yet there are often two that dominate. Still, there is much more opportunity for choice in a system of many parties rather than with few. The Independent Party of the United States has almost no influence currently. Choosing Democrat or Republican typically has little to do with what are supposed to be either of their core values and objectives according to the definition of these words. More often, attention is paid to what is seen as their popular current agendas, track records, and even the stereotypes surrounding the groups. These considerations are of little use when trying to make a carefully weighed decision. Too often people have already decided on the answer before they hear the question. Many times the fine points of both groups are one and the same, yielding no real choice. Of what value is democracy in this situation? One could just as easily make a right or wrong turn as the other when their methods are similar or even identical. What is worse still is that valid concerns go unaddressed and the big picture is narrowed by this lack of perspective. Also, the parties could unite on issues more readily if they were persuaded to do so.

Times could get much worse in any single country while the world as a whole continues to inch toward Utopia. Assuredly many will fall by the wayside, as many already have, on this long, hard road to prosperity. Drastic changes may become necessary in a nation. To make any of these high-level changes correctly, a degree of unity will be required, but also, drastic changes must be faced with as much caution and forethought as possible. If a situation begins to develop, start planning for all possible outcomes so that we are not caught completely by surprise.

Here are some thoughts on how to best initiate change in the system:

- Unite—The people who agree on any proposed modifications to the system must unite to effectively voice their concerns. Perhaps you don't agree with all of the issues, but you think the good would outweigh the bad or that certain basics can be agreed upon. Regardless, there is some certain number of people required who agree on some number of obvious changes that will bring betterment before step two can be approached. Develop your own list of changes you think will be most beneficial and try to justify them. Even if you agree with none of my personal proposals, the most important feature of step one is to be as specific as possible about the propositions. Vague conceptions of direction have tended to spiral out of control in the past. There are numerous examples throughout history, treaties, supposedly aiming for unity, that have led to full scale war, movements that began in the name of love and ended in mass suicide, the defense of national cultures that somehow become genocide along the way, and the intentionally beneficial measures which, in practice, do more harm than good. All are instances of poor planning and veiled intentions which would have been ferreted out by proper clarification of what is to be done, why it is to be done, and what monitoring procedures are to be kept on the success or failure of the intended results. Keep the specific goals clearly in mind, or they may become something horrible. The how is as important as the what and the why. The tactic is every bit as relevant as the strategy.

- Collaborate—Once united, the group must pool its resources. They should find the professionals among them to provide effective guidance. The scientists can discuss the scientific problems and opportunities along with how best to approach them. The teachers' collaborations should share their most effective educational techniques and thoughts for building a better future. So on and so forth, the entire group must communicate, offering opinions and ideas on every aspect of the coming Utopia's development. The group must decide who among them would make the best political figures and then push to get them elected. Consider the qualifications carefully. These people must strive to fulfill the group's objectives without distraction and be willing to be held accountable for their position of representation. These two steps are the basic outline for implementing high-level changes. Unite and collaborate, and do both of these things effectively and efficiently.

- Expect Results from the Government—Say, for instance, the group becomes large enough to get one of their own elected as president. Many would say, once elected, this person could get few of the desired objectives accomplished because of the current establishment's influence over the legislature. They would say Congress would fight against the issues. This representative should state at the outset that they want no votes from people who do not agree on the base agendas. They must insist on the cooperation of the voters to see the goals achieved. So let's consider that they have managed to gain the presidency, but still no progress is made toward the group objectives. Every person in the group must then use his or her vote and voice to assure their representatives that none of them will ever be voted for again if they stand in the way of what the group seeks. This would be a constructive step in the direction of attaining the vital changes once they are identified. A single office alone will not give the group what it wants. If the system is to be refined, it will require efforts from both society and its representatives. The collective must research, observe, and act constructively to gain the proper objectives. Thankfully, one representative cannot enact drastic change. We need an army of representatives who are fiercely devoted to betterment and who will not be swayed from their course by corruption, but we also must seek to use our own influence to encourage the system to give responsibility where responsibility is due. Never expect perfection or unreasonable performance, but never allow corruption or unnecessary complacency to go unannounced.

- Expect Results from Yourselves—If highly effective representation were ever achieved, the onus would be more on the citizens, than on the government to build Utopia if there is to be any degree of self-rule at all. We should never look to the government to solve all of our problems for us. We should only expect them to perform their specified functions as well as can be expected and to encourage them to exceed all expectation. This is the same attitude they should take of us. We should work hand in hand with our governments to reshape the world for the better. Remember that whatever your view of Utopia may be, we must have an appreciation for structure, justice, law, and the previously mentioned elementary hallmarks of its being in order to achieve our potential. Society must have some degree of responsibility if anyone's view of the ideal is to be achieved.

- Support and Defend Positive Achievement—If an entity is interfering with the high-level issues in a detrimental way, such as lobbying for, or outright bribing representatives for, some change that hinders the obviously correct path, it must be first identified, and then persuaded to cease its efforts. One way to make this happen is to pursue the punishment of those persuaded against the represented individuals' chief concerns. Another is for the group to see the business entity as a target for legal attacks. These attacks consist of means such as no members supporting the entity in any way, spreading the information of the entity's corruption, thus damaging their reputation, voting with their dollars against the entity's interests, and by monitoring and scrutinizing their actions to a higher degree (since there has become a reason to do so). Even if it means paying more for a substitute product, we can affect the economy so long as no monopolies exist. This would discourage any attempts to harm the group's objectives for personal gain. It would also be a constructive initiative for citizens to take in defense of their values. Society must be active in its attempts to build and maintain the ideal state, or it can never exist.

- Face Major Obstacles Constructively—As previously stated, in case of serious problems (which are constantly prophesied), violence should be seen as a last resort. In case of a near collapse or an eminent revolution, consider that if a government were to get out of control and start harming a country severely, the masses could stop paying taxes as opposed to inciting violence. If a significant percentage of the country were to refuse to pay any taxes until certain critical demands were met (obviously beneficial changes the majority decides are crucially and immediately required) a large portion of the remaining population would stop paying theirs as well, just to get a perceived free ride. The government cannot jail the masses beyond a certain point. Any action, including nonaction, in dire straits could be detrimental or even fatal, but if things are out of control, some action may be required. If so, a near universal refusal to pay the government any money would weaken the country overall and perhaps to the point of destruction, but could possibly be a positive recourse to outright rebellion or violence. The sole reason for this observation is that being constructive is always preferable to violence. Consider all available options before choosing your course.

These are merely suggestions on ways to exercise our right to an effective government, ways to see high-level changes made constructively. Many of us are quite lucky to have systems of such value and stability as the US and many others possess and should hope that drastic measures are never called for, but not all countries have the benefits and stability we enjoy. Some face the very real possibility of collapse on a regular basis. Governments are supposed to represent society, not dominate it. It is our responsibility to help them function, but it is also their responsibility to function properly. We should want to see them both achieve their potential to the fullest.

Low-Level Adjustments
    More than the high-level adjustments will be required. A society could exist with the greatest possible structure (whatever that may be) and still fail to live up to its potential if the people's aims and everyday mentalities were focused inappropriately. Society has a responsibility to excel at its endeavors just as the government faces its own challenges. How do the views and attitudes of Utopians differ from our own? If we grew up, spent our lives, and died in the ideal environment, we could see more clearly how our current attitudes vary from the ideal. What bad experiences in your life could have been prevented by an improved overall environment? Home life, school life, professional life, private life, public life, how would better environments have improved your current situation? How would they have changed your outlook? In imagining a better world, what do you see as the intrinsic features of the average person's attitude and mind-set? Would not they all have the attitude of studying and learning from the past instead of morbidly dwelling on it? Wouldn't they all seek to improve the aforementioned aspects of their environment? Wouldn't they seek to turn the hands of the clock forward rather than backward? Wouldn't they value their lives, treasuring the sparse, fleeting time we all possess? Utopia could not exist without these attitudes. There is a relationship between our attitudes, characters, and environments. They can chart our progress as clearly

as any other factor. A society rife with these attitudes would refine faster. It would, to a greater extent, seek to improve its environments. Conversely, better environments and structure would encourage people to adopt better attitudes, take more personal initiative, and have a greater appreciation for their advantages. One encourages and the other.

A final thought on the mind-set required to find Utopia: one of the appropriate attitudes a successful society possesses regarding itself is an appreciation for beneficial characteristics. If we have something of value in society, if anything is to be accomplished, then it should be worth defending. To illustrate, the world is going to end tomorrow. What do you do with your last day on Earth? Many get this question wrong. The correct answer is, try to prevent the world from ending. Even if the attempts appear futile, team up and try to save what is obviously worth saving. Go down fighting. If you are forced to make a final statement, let these actions be your testament. If you were to do anything else, what would that say about your values? In a similar vein, why are we here? The answer is, to make the world a better place and to answer those kinds of questions. Why not something else? Because we can make the world a better or worse place and because your asking the question of why we are here demonstrates very well the desire for betterment. Curiosity and wonder breed a desire for understanding. A desire for understanding breeds answers. Answers breed capabilities and options. Once these are gained, it is far more natural for us to want to put them to good use in constructing, rather than squandering all that we've gained, and especially so when considering what has been built thus far.

2084

So what expectations should we set for ourselves? What will be considered satisfactory performance? George Orwell suggested in 1949 that by 1984 we could have a nightmare world for a reality. Who knows, perhaps his warnings of the dangers of excessive control have helped to prevent just such a world from forming. Perhaps many learned a lesson from his dire abstraction. Well, by 2084, we could be much closer to Utopia than we may think. Could it happen, in my lifetime, that society will close in on an ideal state? It is possible, though highly unlikely. It is not currently seen as a goal, and this is its greatest hindrance. It is doubtful that I will live to see the world unite and pursue betterment as priority number one. Its inevitable yet slow progression should only serve to motivate us further, rather than to discourage our efforts. The ideal society may be thousands of years away, especially when considering both the potential and conceivable setbacks and our sometimes unhelpful reactions to them. Note, please, that whatever the estimated range of this goal should be, it must never become an excuse to tolerate mediocrity or to encourage hopelessness.

The amount of progress we should expect is for society to decide. We should at least have a somewhat clearer picture of Utopia by 2084, one that is generally accepted. Certain obvious maleficent attitudes, mind-sets, and problems should be faded and certain gains achieved. Research to find what advancements have been made, and decide for yourself if we are in a slump or if we are currently accelerating. Do not ignore the setbacks, but take them in due consideration. What have we learned from them? Was something missed? What basic public knowledge and goals still require attention? Where is betterment being most constrained? Where is it most thriving? Ask yourself what is hindering and strengthening it in every situation you encounter.

There are assuredly those who will call this optimistic perspective naive. Yet, being based on evidence, it is not specifically optimistic. It is realistic. The reality of the situation happens to prove optimistic rather than hopeless. There are those who will say the world isn't ready for admittedly positive changes that will be proposed and encountered. They will say this because either they feel they are superior and the average person is a Neanderthal or because a thief thinks everyone else steals. They are the ones who are naive and who are not ready for change, although they could become ready if they weren't so concerned with stagnation as opposed to betterment. They fail to appreciate the full scope of such phenomena as self-sacrifice that our species demonstrably exhibits. They are deluding themselves because they believe more in our capacity for failure than our instinct for betterment, which has built the tangible, visible world in which we live, and cannot grasp the awesome power of unity and construction. It is a power too often underappreciated. Instances of its results are not apparent to the mind-set that focuses on negativity. They are not ready for these changes, because they are afraid. They fear a disaster brought on by change. It is irrational to suppose that without the threat of a prison sentence, people will mass toward hard drugs and destroy society. It is irrational to believe that humanity is not capable of working together as a whole, that global unification spells certain doom. We should give society credit when credit is due, and there is much to be found. Most predictions involving a one-world government are irrationally dire. How can you imagine a better world without unity playing a highly involved role? The evidence is everywhere.

Concerning the liberal versus conservative points of view, look to their definitions. "Liberal" implies liberty and maximum capacity for change. "Conservative" implies maintenance of the beneficial and minimization of change. Though many of my propositions would label this work as being extremely liberal, I think we are somewhere in the general vicinity of the halfway point of reaching Utopia (depending on how you measure it), and so there is very much that should be conserved. We must never cause change simply for the sake of change. We should value absolute betterment above absolute liberation or absolute conservation. If we change compulsively, we abandon the positive once found. If we stagnate compulsively, we will never achieve any of our potential. Knowing where to change and where to stand fast is the challenge. Maximize liberty and government while valuing the development of society.

The idea of building a better world is not a new concept. Millions have already contributed to its foundation. Since the trend thus far in human history is that we eventually do things the smart way despite the factors of change resistance, power struggles, destructive traditions, and historic feuds, just to name a few, it stands to reason that in time we will learn to cooperate collectively and build a better world. Since continual improvement is an integral part of our composition, anything that can possibly go right, eventually will go right. There is nothing else to work for of comparable worth to this incredible aspect of humanity.

Utopia will likely arrive in time, but we can speed up or slow down its progress. It may be destined to arrive after several thousands of years, or it may get here sooner than any of us think possible. It cannot build itself, however. There must be those who want to improve our world. There must be those who consciously pursue continual progression on every level of their lives. There will always be disagreements over which is the right path to take, but the rule should be to always seek improvement, keep this goal in mind. We must never resort to conflict as an answer to anything that may have another solution. We must never allow a disagreement over the best way to proceed to become anything other than a problem to be solved as a team. Debatable points should be exercises in deliberation, in problem solving. Complacency is to be avoided, while creativity and thorough, grinding study embraced. Any step can be redrawn, but idleness to such problems as we face is inexcusable. Lives depend on getting the answers right. We will be one step closer to our goal so long as we keep in mind that our social experiment, which has been going on since before recorded history, has already made several wrong turns that have been righted. We are much nearer still if we recognize the value of documenting our errors in great detail so as to learn from them and cease repeating them. There must be some room for careful experimentation if there is to be any progress. Change must only be made when attempting to gain a clear positive advantage with all possible repercussions in mind and taken into full consideration.

To summarize this book's key points, look to the future, try to visualize where society is going over the stretch of time, and speed its progression. Picture our coming Utopia. Picture peace and unity. Picture blatant corruption as being a thing of the past, never dared to be practiced in the face of a more effective system of justice. Picture a world where war has become a distant, unpleasant memory, disappearing along with famine as an obvious enemy of all mankind. Universal peace has been dreamt of (literally) for centuries, often by those who experience firsthand the realities of conflict and discord. Your vision of this better world accelerates its arrival. Your open-minded pursuit of its most efficient and effective progression makes its own difference in society.

Again, we must always ask ourselves in regard to each and every political and social decision, "Is this going to take us closer to where we want to go?" We must focus on the destination and take positive action to achieve it if we are to speed its arrival. This destination should be to continually seek betterment while keeping an eye on the overall situation. The whole world can become a Utopia. In fact, for Utopia to exist, it must consist of a unified world. It is waiting to be built. It is currently rising, as it always has been, and people the world over should be doing all that can possibly be done to facilitate its arrival or to at least stop interfering with its progression. It cannot get here fast enough. Nothing else can bring the kind of fulfillment striving for our potential can bring. Let's find out how far humans can progress. Let's find the extent of our capabilities and test our capacity for success.

To accurately picture your own version of Utopia, consider all proposals, the realities of human strengths and weaknesses, consider the needs and desires of our species, and apply this to our potentiality as a society. Construct your personal blueprints for the ideal society with extreme care. Slight errors in theory can incur heavy consequences in application. Remember that humankind has been striving for these goals since before the discovery of the usefulness of recording information. Consider the concepts and proposals of others. Consider today's needs as well as the long-range objectives. Massed creativity and scrutiny will be required to realize and achieve our objectives quickly. Once the objectives are formed, we must consider how to practically reach them if they are to be put to use. Imagine the prize of achievement, the lives we may save, the prosperity and peace we may maintain, the wonders that await our search. Picture Utopia's beauty, the beauty of gaining a better understanding of our potential, and you can become a part of it.

The Utopian View in Summary

The Core Theory

1. Humans tend to progress over time.
2. Short of extinction, this progression will eventually lead to the creation of an idyllic society.
3. Given these two points are facts, the appropriate action is to hasten this progression by discovering the ideals of life and to adopt principles that will benefit this cause.

Society's Hallmarks

- Civility
- Safety
- Prosperity
- Knowledge
- Health
- Technology
- Independence
- Comfort
- Efficiency in Dealing with Social Issues
- The Preoccupations of the Collective Will

The Utopian Precepts

- Value betterment.
- Defend all worth defending.
- Value justice and law. Respect and appreciate their centers of operation.

- Value freedom and responsibility. Accept the consequences of your actions as well as their rewards.
- Value safety. Constantly search for ways to promote it.
- Achieve unity through peaceful means. (Violence should always be a final recourse.)
- Aim for and seek out a beautifully designed future.
- Consider our physical environment, both natural and man-made, as sacred.
- Question established standards using facts and logic. Nothing is beyond question.
- Establish your own principles. Base them on honorable goals.
- If encountered, undermine tyranny. Help the system of justice to function as well as it can.
- Live your life in such a way that you can smile as it ends.
- Seek to gain as much knowledge as you can from life. Continue learning even while on your deathbed.
- Consider individuals equally and as potential comrades until they give you reason to either respect them or hold them in contempt. Be cautious and wary, not paranoid and prejudiced. Beware of forming assumptions based on appearances.
- Take pride in accomplishment, not heritage, chance, or subjectively aesthetic personal features.
- Never allow tradition to interfere with progress. Never allow feeling to hinder thought. Never allow instinct to unnecessarily influence judgment.
- Solve life's mysteries, and explore the unknown.

Note that the hallmarks and precepts are in no specific order of priority or absolute form as yet. Feel free to amend and prioritize them for yourself. There are assuredly many more to add to this short list. Only evidence, facts and logic can formulate them, and only the test of time (field testing) can prove their worth. Remember, too, that in reflection we already have much history to consider them against.